Plasticity and Morphology
of the
Central Nervous System

Plasticity and Morphology of the Central Nervous System

Edited by

C.L. Cazzullo, E. Sacchetti, G. Conte, G. Invernizzi and A. Vita

Istituto di Clinica Psichiatrica
Università degli Studi di Milano
Milan, Italy

KLUWER ACADEMIC PUBLISHERS
DORDRECHT / BOSTON / LONDON

Distributors

for the United States and Canada: Kluwer Academic Publishers, PO Box 358, Accord Station, Hingham, MA 02018-0358, USA
for all other countries: Kluwer Academic Publishers Group, Distribution Centre, PO Box 322, 3300 AH Dordrecht, The Netherlands

British Library Cataloguing in Publication Data

Plasticity and morphology of the central nervous system.
  1. Man. Central nervous system. Physiology
  I. Cazzullo, C.L. (Cairo Lorenzo)
  612'.82

ISBN-13:978-94-010-6870-3   e-ISBN-13:978-94-009-0851-2
DOI: 10.1007/978-94-009-0851-2

Published in the United Kingdom by Kluwer Academic Publishers,
PO Box 55, Lancaster, UK.

Kluwer Academic Publishers BV incorporates the publishing programmes of D. Reidel, Martinus Nijhoff, Dr W. Junk and MTP Press.

Lasertypeset by Martin Lister Publishing Services, Carnforth, Lancs.

Contents

Preface   viii

List of Contributors   ix

SECTION 1: SCHIZOPHRENIA: THE IMPACT OF NEUROIMAGING TECHNIQUES

1. Structural and developmental abnormalities in schizophrenia
   assessed with MRI
   *N.C. Andreasen, J.C. Ehrhardt, V.W. Swayze II, W.T.C. Yuh, N.
   Blakley and S. Ziebell*   3

2. Brain morphology in schizophrenia by magnetic resonance imaging
   *A. Rossi, P. Stratta, V. Di Michele, M. Gallucci, A. Splendiani, R.
   Passariello and M. Casacchia*   13

3. Brain density patterns in schizophrenia
   *A. Vita, C. Sina, E. Sacchetti, G. Valvassori, A. Calzeroni, G.M.
   Giobbio, G. Invernizzi, V. Bernasconi and C.L. Cazzullo*   21

4. Lateral ventricular size and clinical features in schizophrenia
   *C. Maggini, M. Guazzelli, S. Starnini, M. Ardito, L. Rosi, F. Busoni
   and E. Camerini*   33

5. Hippocampal metabolic function in schizophrenia
   C.A. Tamminga, K. Lyons, S.K. Kuo and G.K. Thaker   43

SECTION 2: AFFECTIVE DISORDERS: THE IMPACT OF NEUROIMAGING TECHNIQUES

6. CT and MRI findings in affective disorders: clinical and research
   implications
   H.A. Nasrallah, J.A. Coffman and S.C. Olson   53

7. Neuromorphological correlates of mood disorders: focus on
   cerebral ventricular enlargement
   *E. Sacchetti, A. Vita, A. Calzeroni, G. Conte, F. Pollastro, A. Terzi,
   G. Valvassori, G. Invernizzi and C.L. Cazzullo*   63

8. Brain imaging with positron emission tomography in affective
   illness
   M.S. Buschsbaum and J.C. Wu   73

9. The impact of starvation on brain morphology and function in
    eating disorders
    W. Schreiber, C. Laver, K.M. Pirke, H.M. Emrich, G. Leinsinger,
    E.A. Moser and J.C. Krieg   83

SECTION 3: THE IMPACT OF NEUROPHYSIOLOGICAL STUDIES IN PSYCHIATRY

10. Neurometric subtyping of depressive disorders
    L.S. Prichep, E.R. John, T. Essig-Peppard and K. Alper   95

11. Lateralization patterns of verbal stimuli processing in
    schizophrenia patients
    S. Galderisi, A. Mucci, M. Maj and D. Kemali   109

12. EEG patterns in schizophrenia: a familial study
    C. Colombo, O. Gambini, F. Macciardi, M. Locatelli, G. Calabrese
    and S. Scarone   117

13. Some aspects of psychophysiological studies in psychiatry
    J. Saarma, G. Morozow and N. Zharikov   125

SECTION 4: THE IMPACT OF GENETIC STUDIES IN PSYCHIATRY

14. Genetic marker studies in schizophrenia
    L.E. DeLisi   133

15. Amish study of genetics of affective disorders
    J.A. Egeland   147

16. Affective disorders segregation structures according to clinical and
    pharmacological features
    E. Smeraldi, M. Gasperini, F. Macciardi, A. Orsini, M. Provenza, G.
    Sciuto and C. Bussoleni   159

17. Psychiatric disorders in families of phenylketonuric patients
    E. Smeraldi, F. Macciardi, M. Leporatti, O. Gambini, A. Orsini, C.
    Fara, R. Valsasina and E. Riva   169

SECTION 5: PLASTICITY AND VULNERABILITY OF THE CNS IN PSYCHIATRIC DISORDERS

18. Neuronal plasticity in the central nervous system: a
    pharmacological approach
    A. Consolazione   177

19. A neurodevelopmental perspective on some epiphenomena of schizophrenia
   R.M. Murray, M.J. Owen, R. Goodman and S.W. Lewis 185

20. Clinical investigations of plasma homovanillic acid concentrations
   M. Davidson, R. Kaminsky, S. Jaff, N. Runyon and K.L. Davis 203

21. Glucose-6-phosphate dehydrogenase deficiency and psychoses
   A. Bocchetta, M. Del Zompo and G.U. Corsini 211

22. Early separation events and noradrenergic status during major depression
   G. Conte, A. Calzeroni, A. Pennati, A. Terzi, A. Vita and E. Sacchetti 221

23. Vulnerability and plasticity of monoamine neurotransmitter systems in affective and personality disorders
   L.J. Siever, E.F. Coccaro, M. Davidson, L. Howard, L. Harter and K.L. Davis 231

24. Concluding remarks: plasticity and morphology of the central nervous system in schizophrenia and affective disorders
   C.L. Cazzullo 251

Preface

In the last few years ways of thinking in psychiatry have undergone considerable change thanks to advances in the fields of morphology and plasticity of the CNS, particularly with regard to schizophrenic and mood disorders. In addition, the rapid and considerable development of neuroimaging techniques (CT, MRI, PET and computerized EEG) and of molecular genetics (through DNA recombinant methodologies) have widened the approach to these disorders in a way unimagined a few years ago.

These advances and the new etiopathogenetic hypotheses that have sprung from them were the central theme of the Second International Meeting on Schizophrenia "Morphology and Plasticity of the Central Nervous System – A Challenge for Psychiatry of the Nineties" which was organized by the Association for Research on Schizophrenia (ARS), the Schizophrenia Research Center of the Institute of Psychiatry of the University of Milan and the T. and F. Legrenzi Foundation, held in Milan on October 22–24, 1987.

This book contains the contributions from participants of the meeting, which took place in a warm and friendly atmosphere and marked by lively and exhaustive discussions on the various papers. The contributions were recently revised for the present publication. We would like to express our appreciation to the book's contributors for the high quality of their reports.

The Editors

List of Contributors

N.C. ANDREASEN
The University of Iowa Hospitals and Clinics
Psychiatric Hospital
500 Newton Road
Iowa City, IA 52242
USA

A. BOCCHETTA
Department of Neurosciences
Clinical Pharmacology
University of Cagliari
Cagliari
Italy

M.S. BUCHSBAUM
Department of Psychiatry and Human
    Behavior
California College of Medicine
University of California, Irvine
Irvine, CA 92717
USA

C.L. CAZZULLO
Department of Psychiatry
University of Milan
School of Medicine
via F. Sforza, 35
20122 Milan
Italy

C. COLOMBO
Department of Psychiatry
University of Milan
School of Medicine
via F. Sforza, 35
20122 Milan
Italy

A. CONSOLAZIONE
FIDIA Research Laboratories
Abano Terme, PD
Italy

G. CONTE
Department of Psychiatry
University of Milan
School of Medicine
via F. Sforza, 35
20122 Milan
Italy

M. DAVIDSON
Department of Psychiatry
Bronx VA Medical Center
130 W. Kingsbridge Road
Bronx, NY 10468
USA

L.E. DE LISI
Department of Psychiatry
School of Medicine
Health Sciences Center
SUNY at Stony Brook
NY 11794
USA

J.A. EGELAND
Amish Study
North Research Division
49 Sylvania Road
Hershey, PA 17033
USA

S. GALDERISI
Department of Medical Psychology and
    Psychiatry
First Medical School
University of Naples
Naples
Italy

C. MAGGINI
Head, 3rd Psychiatry Clinic
University of Bologna
Bologna
Italy

R.M. MURRAY
University of London
Institute of Psychiatry
De Crespigny Park
Denmark Hill
London SE5 8AF
UK

H.A. NASRALLAH
The Ohio State University
Department of Psychiatry
Room 071, Upham Hall
473 West 12th Avenue
Columbuus, OH 43210-1288
USA

L.S. PRICHEP
Brain Research Laboratories
New York University Medical Center
New York, NY 10016
USA

A. ROSSI
Department of Internal Medicine
Chair of Clinical Psychiatry
University of L'Aquila
L'Aquila
Italy

Y.M. SAARMA
Director of the Institute
Academician of the USSR
Academy of Medical Sciences
1199034 Moscow
USSR

E. SACCHETTI
Department of Psychiatry
S. Paolo Hospital
via Di Rudini' 8
20142 Milan
Italy

W. SCHREIBER
Maz-Planck Institute for Psychiatry
Kraepelinstrasse 10
8000 Munich 40
Federal Republic of Germany

L. SIEVER
Bronx VA Medical Center
130 W. Knightsbridge Road
Bronx, NY 10468
USA

E. SMERALDI
Department of Psychiatry
Head, 3rd Psychiatry Clinic
University of Milan
Milan
Italy

C.A. TAMMINGA
Maryland Psychiatric Research Center
PO Box 21247
Baltimore, MD 21228
USA

A. VITA
Department of Psychiatry
University of Milan
School of Medicine
via F. Sforza, 35
20122 Milan
Italy

SECTION 1:
SCHIZOPHRENIA: THE IMPACT
OF NEUROIMAGING
TECHNIQUES

1
Structural and developmental abnormalities in schizophrenia assessed with MRI

N.C. Andreasen, J.C. Ehrhardt, V.W. Swayze II, W.T.C. Yuh, N. Blakley and S. Ziebell

INTRODUCTION

Our current MRI studies grow out of earlier pilot work. In our first MRI study, we examined a sample of 38 schizophrenics and 45 normal controls using principally a midsagittal cut for our analyses. In these studies, we focused our emphasis on morphometrics, or the study of structural size that had potential functional significance. In this first study, we observed a significant decrease in frontal lobe size, cerebral size, and (somewhat to our surprise) cranial size [1]. We were unable to explain this finding on the basis of differences between patients and normal controls and possible confounding variables such as sex, height, or weight. In that study, however, it did appear that there was a possible sex effect in the schizophrenic patients, with slightly more males showing the frontal, cerebral, and cranial abnormalities. These findings were relatively striking. 39 percent of the schizophrenic patients had frontal lobe size outside the control range, while percentages for cerebral and cranial size outside the control range were 25 and 18 percent, respectively. In this study, it also appeared that there were abnormalities in callosal shape in the schizophrenic patients, with a reversal of the "normal" sexual dimorphism that has been observed by other investigators [2]. In this study the female schizophrenic patients had smaller splenia than did the males [3].

This first study had a number of limitations. Because it was done with limited funding, we were unable to obtain a full range of scanning sequences on both patients and controls. Consequently, the morphometric analyses were limited to a midsagittal cut, which was available for both groups. Other structures of interest, best seen on coronal cuts, such as the amygdala and hippocampus, could not be measured. Further, the controls were selected by recruiting hospital personnel who were not closely matched to the patients in education or sociodemographic status.

We are now in the midst of completing a second MRI study that improves substantially over the methodology of the first study. This second study has several goals. First, we are seeking to obtain detailed morphometric measures of brain structures in all three dimensions. Second, we are attempting to determine whether increased white matter lucencies (unidentified bright objects or "UBOs") occur with greater frequency among psychiatric patients. Third, we are attempting to develop reliable methods for measuring T1 and T2 relaxation times and to determine whether they are abnormal in patients with major psychoses. Fourth, we are evaluating the specificity of our findings by examining both schizophrenic patients and manic patients. Of course, if we observe abnormalities we are interested in correlating them with clinical phenomena, such as the nature of the symptomatology or the presence of tardive dyskinesia.

Data collection for this study is nearing completion, but data analysis is not yet complete. In this paper, we will report some preliminary findings from our morphometric studies and our studies of white matter lucencies. The characteristics of the sample currently collected are as follows:

TABLE 1. Characteristics of sample collected to date

	Schizophrenics		Bipolars		Controls	
	Males (N=33)	Females (N=18)	Males (N=27)	Females (N=19)	Males (N=25)	Females (N=20)
Age	32.6	35.4	33.1	34.6	32.2	37.3
Height (cm)	174.0	164.9	176.9	161.7	176.8	165.0
Weight (kg)	77.0	64.3	81.5	69.2	76.1	63.9

In studying this sample, we obtained eight coronal slices, centering the sequence on the optic chiasm and obtaining three cuts anterior to this and four cuts posterior using a 1 cm slice thickness. These coronal images were obtained using the inversion recovery mode on our .5 Tesla scanner with TI equal to 600 ms and TR equal to 1700 ms. We obtained five sagittal cuts centered on the midsagittal region using 1 cm slice thickness and a similar inversion recovery sequence. In addition, we obtained four to eight spin-echo images in the transverse plane with a TE of 40 ms and TR of 500 ms beginning at the mid-cerebellum and ending above the body of the ventricles. We selected this particular group of sequences in order to be able to use the sagittal and coronal planes for anatomic morphometric evaluation and the transverse cut to look for small focal lesions that might reflect patchy gliosis that has been reported in schizophrenia by some neuropathologists [4].

One major goal of this study was to confirm our previous findings from MRI Study 1. In this second study, we obtained quite different results, however. The results of the two studies are compared in Table 2. The differences between male subjects and controls in Study 1 were highly significant, while no differences were observed in MRI Study 2. Sex, height, and weight corrections do not change these results. In general, both groups appear to have regressed to the mean in Study 2, and the controls in Study 2 have distinctly smaller cerebral and cranial measures. Table 3 shows statistical data comparing schizophrenic patients from MRI Study 2 with controls from MRI Study 1 with a one-tailed t-test. When this comparison is made, the schizophrenics continue to differ from the controls in frontal area with a trend toward a difference in cerebral area, but no significant difference in cranial area. This may suggest that part of the failure to replicate our previous findings results from a change in the control group. Controls in MRI Study 2 were closely matched to the patients for education and sociodemographic status. Thus, they were relatively socio-economically deprived compared to the general population. We conclude from this that one possible explanation for our nonreplication is the change in the nature of the control group. The controls in MRI Study 2 may have suffered social and environmental handicaps early in life that limited their cerebral development in a manner similar to that possibly occurring in patients with schizophrenia.

TABLE 2. Midsagittal measures (in square centimeters) for male subjects

MRI STUDY 1

	Schizophrenics (N=26)		Controls (N=27)			
	Mean	SD	Mean	SD	t	p
Frontal area	50.57	3.99	54.92	5.37	3.36	.0015
Cerebral area	86.72	8.08	95.22	9.05	3.61	.0007
Cranial area	152.58	9.78	164.33	11.51	4.01	.0002

MRI STUDY 2

	Schizophrenics (N=26)		Controls (N=27)			
	Mean	SD	Mean	SD	t	p
Frontal area	51.95	5.59	51.14	4.56	.544	NS
Cerebral area	91.55	9.49	88.86	7.80	1.052	NS
Cranial area	163.04	14.27	160.02	9.46	.847	NS

TABLE 3.   MRI Study 1 vs. MRI Study 2

| | Schizophrenics MRI Study 2 | | Controls MRI Study 1 | | | |
	Mean	SD	Mean	SD	t	p
Frontal area	51.95	5.59	54.92	5.37	1.44	.03
Cerebral area	91.55	9.49	95.22	9.05	1.41	.08
Cranial area	163.04	14.27	164.33	11.52	.36	.56

In addition to exploring for abnormalities in the frontal system using morphometric measurements, we were also interested in determining whether structural abnormalities could be observed in several other brain regions including the basal ganglia, the amygdala and hippocampus, and the corpus callosum. Several previous neuropathological reports have indicated decreased size in the hippocampus and parahippocampal gyrus [5,6,4,7,8]. We postulated that we might be able to measure atrophy in these regions. Since these structures are so small, and since our slice thickness was only 1 cm, we were essentially limited to area measurements. Although these structures can sometimes be seen on several cuts, and we did measure them whenever present, nevertheless we have chosen to base our current report on the area of the relevant structure on the cut on which it has the largest size (since this is probably the most valid measurement). We examined both absolute measurements of size and ratio measurements, using the area of the subcortical structures and the area of the brain seen on the same cut. With absolute measurements we found no differences between schizophrenics and controls. Using ratio measurements, however, some differences did emerge. These appear in Table 4. No differences are seen in the amygdala or hippocampus, and consequently in our sample using MRI we are unable to confirm atrophic changes in these regions in schizophrenia. Somewhat strangely, increases are seen in the caudate bilaterally and in the putamen on the left side.

TABLE 4.   Basal ganglia and limbic structures in schizophrenics
and normals*

	Schizophrenics (N=51)		Controls (N=45)			
	Mean	SD	Mean	SD	t	p
Caudate						
Left	1.72	.41	1.50	.31	2.21	.03
Right	1.86	.40	1.70	.34	2.21	.03
Putamen						
Left	4.20	.53	3.95	.57	2.24	.03
Right	4.08	.52	3.88	.69	1.56	NS
Amygdala						
Left	2.66	.59	2.67	.48	.11	NS
Right	2.66	.52	2.65	.55	.11	NS
Hippocampus						
Left	1.24	.46	1.32	.36	1.12	NS
Right	1.29	.38	1.34	.37	.77	NS

*   Based on a ratio of the specific structure to the brain area, as
seen in the coronal cuts

We piloted a number of different methods for measuring the
corpus callosum, and we finally elected to use a method that
divides the callosum into fifths, or quintiles. We selected this
method because it appeared to provide the most sensitive
approach to identifying possible regional differences in the
callosum, such as increased area in either the anterior callosum
or the splenium. Previous studies using postmortem brains have
indicated structural abnormalities in the callosum in relation to a
variety of conditions. Witelson [9] reported that the callosum is
larger in left-handers, probably reflecting increased fibers and
increased interconnections between the two hemispheres in
left-handers. De Lacoste-Utamsing et al. [2] reported
differences in callosal shape in males and females, with females
having an increase in callosal size in the splenium, again
assumed to reflect and increase in the number of fibers. De
Lacoste-Utamsing et al. interpret this to mean that females have
less differentiated brains than do males, a finding sometimes
referred to as "sexual dimorphism." In our sample we had a
relatively small number of left-handers (N = 7) in the normals.
In this small sample we did not find the callosum to be larger in
left-handers as Witelson did. We also did not observe any
sexual dimorphism in our normal patients. There were no
significant differences between males and females in callosal size
in any of the five quintiles. These data are shown in Table 5.

TABLE 5.   Does sexual dimorphism occur?

	Male Controls N=22	Female Controls N=18	p
Callosal:cerebral ratio	7.66	7.80	NS
Anterior fifth	26.66	27.27	NS
Mid-anterior fifth	15.53	15.72	NS
Middle fifth	14.11	13.81	NS
Mid-posterior fifth	15.67	17.42	NS
Posterior fifth	28.12	26.08	NS

We were also unable to confirm any reversal in sexual dimorphism in schizophrenia. The schizophrenic males did have a significantly larger anterior fifth callosal area when both left- and right-handed males were pooled. This difference disappeared, however, when we examined only right-handed males. This finding is, in any case, opposite to that reported earlier by Nasrallah et al. [3] in our first study, in that the increase in size is anterior and is in the males.

We have also explored ventricular size in some detail. In these studies we have rather consistently confirmed previous work. Schizophrenics as a group have greater ventricular size throughout the entire ventricular system than do the control subjects. This finding emerges clearly in both sagittal and coronal cuts. Further, in our earlier CT work, we found that most of the increased ventricular size is contributed by the male patients. These findings are summarized in Tables 6, 7, and 8. Bipolar patients, on the other hand, do not have prominent increases in ventricular size as shown in Table 9. Thus, using the improved resolution of MRI and coronal cuts to reduce partial volume effect, we appear to find that ventricular enlargement is more common in schizophrenia than in bipolar disorder, suggesting that structural abnormalities reflecting either atrophy or dystrophy may be relatively specific to schizophrenic psychosis rather than common to all psychoses.

TABLE 6. VBR on coronal cuts: All schizophrenic subjects

VBR	N	Group	Mean	SD	t	p
Cut -2	15	Controls	1.14	0.84	1.51	.1428
	22	Schizophrenics	1.93	2.23		
Cut -1	45	Controls	1.55	0.87	2.34	.0221
	48	Schizophrenics	2.21	1.74		
Cut 0	46	Controls	1.90	0.66	2.49	.0149
	51	Schizophrenics	2.46	1.43		
Cut +1	46	Controls	1.65	0.56	3.31	.0015
	51	Schizophrenics	2.35	1.38		
Cut +2	46	Controls	1.23	0.52	3.22	.0020
	51	Schizophrenics	1.87	1.30		
Cut +3	45	Controls	1.43	0.74	2.09	.0397
	51	Schizophrenics	1.90	1.40		

TABLE 7. VBR on coronal cuts: Male schizophrenic subjects

VBR	N	Group	Mean	SD	t	p
Cut -2	5	Controls	0.64	0.33	2.35	.0315
	16	Schizophrenics	2.16	2.51		
Cut -1	25	Controls	1.42	0.74	2.48	.0172
	33	Schizophrenics	2.36	1.98		
Cut 0	26	Controls	1.77	0.55	2.83	.0070
	34	Schizophrenics	2.61	1.61		
Cut +1	26	Controls	1.70	0.51	2.71	.0097
	34	Schizophrenics	2.49	1.59		
Cut +2	26	Controls	1.20	0.47	2.89	.0061
	34	Schizophrenics	1.99	1.50		
Cut +3	26	Controls	1.28	0.65	2.58	.0133
	34	Schizophrenics	2.07	1.62		

TABLE 8.   VBR on coronal cuts: Female schizophrenic subjects

VBR	N	Group	Mean	SD	t	p
Cut −2	10	Controls	1.39	0.91	0.16	.8728
	5	Schizophrenics	1.30	1.13		
Cut −1	20	Controls	1.70	1.02	0.54	.5952
	15	Schizophrenics	1.88	0.98		
Cut 0	20	Controls	2.08	0.75	0.25	.8044
	17	Schizophrenics	2.14	0.91		
Cut +1	20	Controls	1.59	0.62	2.07	.0455
	17	Schizophrenics	2.08	0.79		
Cut +2	20	Controls	1.27	0.58	1.60	.1196
	17	Schizophrenics	1.63	0.77		
Cut +3	19	Controls	1.64	0.83	0.26	.7966
	17	Schizophrenics	1.57	0.71		

TABLE 9. VBR on coronal cuts: All bipolar subjects

VBR	N	Group	Mean	SD	t	p
Cut −2	20	Bipolars	0.93	0.69	0.81	.4260
	15	Controls	1.14	0.84		
Cut −1	45	Bipolars	1.74	0.89	1.07	.2896
	45	Controls	1.55	0.87		
Cut 0	46	Bipolars	2.07	0.80	1.07	.2879
	46	Controls	1.90	0.66		
Cut +1	46	Bipolars	1.91	0.78	1.80	.0751
	46	Controls	1.65	0.56		
Cut +2	46	Bipolars	1.42	0.73	1.42	.1603
	46	Controls	1.23	0.52		
Cut +3	46	Bipolars	1.48	0.92	0.26	.7979
	45	Controls	1.43	0.74		

We sought for periventricular lucencies by blindly rating all scans visually. These blind ratings were done by a team of neuroradiologists. The results of these ratings appear in Table 10. As that table indicates, UBOs were relatively uncommon in all these patients, but occurred with slightly greater frequency among the bipolars. They were very uncommon in both the schizophrenics and the normals. Thus there seems little evidence to confirm patchy gliosis in the white matter of the brain in schizophrenia. The reasons for the increased number of UBOs in bipolars is unclear. We are currently examining our data to determine whether there is an increased rate of hypertension among our bipolar patients.

TABLE 10.   Frequency of UBOs in schizophrenics, bipolars, and normals

	N	%
Schizophrenics	4/54	7.4
Bipolars	9/46	19.6
Controls	3/51	5.9
$X^2 = 5.68$, p = .058		

SUMMARY AND CONCLUSIONS

MRI offers a unique opportunity to study brain structure because it provides clear views of the brain in all three planes with higher resolution than has been previously available with any other imaging modality. If structural abnormalities occur in psychoses, or in relation to differences in brain function in normal individuals, MRI provides an excellent method of detecting them. Nevertheless, this study has been surprisingly negative. In a second, much improved study of a relatively large sample of patients and matched normal controls, we have only confirmed an increase in ventricular size in patients suffering from schizophrenia. While this is important, it does not add a great deal to our previous knowledge. We have not been able to confirm atrophic changes in subcortical regions such as the amygdala or hippocampus or abnormalities in callosal shape or size. The absence of positive findings in the study of subcortical regions may reflect the limitations of the method, however. While MRI has excellent resolution in comparison with CT, nevertheless it is not as clear as direct study of the postmortem brain. Further, our slices were relatively thick (1 cm); greater sensitivity would be provided with finer slice thicknesses, as well as volumetric measurements, and we plan to pursue this approach in the future.

REFERENCES

1. Andreasen, NC, Nasrallah, HA, Dunn, V., Olson, SC, Grove, WM, Ehrhardt, JC, Coffman, JA, and Crossett, JHW (1986). Structural abnormalities in the frontal system in schizophrenia: A magnetic resonance imaging study. Arch. Gen. Psychiatry 43:136-144. 2. De Lacoste-Utamsing, C., and Holloway, RL (1982). Sexual dimorphism in the human corpus callosum. Science 216:1431.
3. Nasrallah, HA, Andreasen, NC, and Coffman, JA (1986). A controlled magnetic resonance study of corpus callosum thickness in schizophrenia. Biol. Psychiatry 21:274-282.
4. Stevens, J (1982). Neuropathology of schizophrenia. Arch. Gen. Psychiatry 39:1131-1139.
5. Falkai, P, and Bogerts, B (1986). Cell loss in the hippocampus of schizophrenics. Eur. Arch. Psychiatr. Neurol. Sci. 236:154-161.
6. Bogerts, B, Meertz, E, and Schonfeldt-Bausch, R (1985). Basal ganglia and limbic system pathology of schizophrenia. Arch. Gen. Psychiatry 42:784-791.
7. Roberts, GW, Colter, N, Lofthouse, R, Bogerts, B, Zech, M, and Crow, TJ (1986). Gliosis in schizophrenia: A survey. Biol. Psychiatry 21:1043-1050.
8. Kovelman, JA, and Scheibel, AG (1984). A neurohistological correlate of schizophrenia. Biol. Psychiatry 19:1601-1621.
9. Witelson, SF (1985). The brain connection: The corpus callosum is larger in left-handers. Science 229:665-668.

2
Brain morphology in schizophrenia by magnetic resonance imaging

A. Rossi, P. Stratta, V. Di Michele, M. Gallucci, A. Splendiani, R. Passariello and M. Casacchia

INTRODUCTION

Numerous studies over the past several years have examined brain morphology of schizophrenics using a variety of techniques including pneumoencephalography and computed tomography (CT) [1 - 2]. The most widely reported finding has been the dilatation of the brain ventricles in schizophrenia [2]. Over the last decade there has been increasing interest in the Corpus Callosum (CC) generated both from post-mortem reported CC abnormalities [3 - 4] and from evidence of defective interhemispheric functioning [5 - 6 - 7 - 8].

Recently, Magnetic Resonance (MR), a new technique for brain imaging, has enabled researchers to overcome some of the technical limitations of CT scan such as bone artifact and low tissue contrast. In addition, MR produces images of greater anatomical detail than CT scans, and allows imaging of the brain in multiple planes, such as sagittal and coronal planes which are, respectively, impossible or difficult to acquire with a CT scanner so that researchers can obtain measures of cerebral structures such as frontal lobe, corpus callosum that otherwise could not be obtained. Up until now, the few MR studies which have been carried out have reported an increased callosal thickness in female schizophrenics [9], higher image intensities with no differences in linear and area measurements [10], and a smaller corpus callosum area [11]. An association between negative symptoms and a reduction in frontal lobe has also been reported [12] although more recently Nasrallah [13] reported an unexpected finding, showing that larger cerebral area was associated with a greater severity of negative symptoms.

We undertook a MR controlled study in schizophrenia exploring areas putatively involved in schizophrenia. The hypothesis is that larger Ventricle Brain Ratio

(VBR) reported in the majority of cerebral morphological studies is only one, and perhaps the most measurable brain abnormality and that other cerebral areas could be involved in a subtle brain rearrangement; Corpus Callosum could be involved in such a rearrangement.

METHOD

30 chronic schizophrenic patients (23 male and 7 female) according to DSM III criteria, were studied by means of MR examinations after informed consent was obtained. This sample was drawn from 36 consecutively admitted patients of whom 6 were unable to complete the MR brain scan because head movements. Their mean age was 31.03±7.37SD years. The patients had been ill for 7.60±5.05 years. They were all being treated with low doses of neuroleptics.

Midsagittal and axial MR scans were also obtained on 25 healthy volunteers (18 male and 7 female). Normal controls were chosen among employees and relatives of the hospital staff and were matched for age (within 3 years), sex and time of scan (the same day) with 25 of 30 schizophrenic patients. Their mean age was 31.32±6.21 years. Patients and controls were right-handed. Handedness was determinated by the Edinburgh Inventory [14].

Subjects were excluded if they manifested recent evidence, or had a history of alcohol or drug abuse, or evidence of neurological or physical diseases, morphological insults to the brain, or had surgical metal or electronic implants which could interfere with MR evaluation.

MR examinations were performed by means of an Ansaldo Esatom 5000 scanner operating at a 0.5 Tesla magnetic field. A 26 cm crossed-ellipse head coil was used. Acquisition and reconstruction matrix were 256x256. Firstly, a midsagittal scan (5mm thickness, Spin-Echo 350/30) was taken. Afterwards, 8 axial slices, 10 mm thick, using the Spin-Echo sequence (TR 1800 msec. and TE 30 msec.) were obtained, at 15° to orbitomeatal plane. All linear and area measurements were made by two radiologists (M.G. and R.P.) who were unaware of the diagnosis, using computer-assisted planimetry on the MR video console. The measurements used in subsequent calculations were the average of the findings of the two above mentioned radiologist. Intra- and interrater reliabilities were strong (r=.91 and .89 respectively).

The mid-sagittal slice which gave the clearest outline of the corpus callosum (CC) was used for measurement purposes. The total surface of the cortical

area (CA), CC area was measured in square centimetres.
CC to brain area ratio (CCBR) was expressed as the
ratio of the CC area to that of the cerebral area plus
CCarea X 100.

VBR (Ventricular Brain Ratio) was calculated on
the axial scan as the ratio of the area of the lateral
ventricles at their largest to that of the entire brain
in the section, expressed as a percentage.

The presence of both positive and negative
symptoms of schizophrenia was rated using the Krawiecka
et al. [15] scale. The scores for expressed delusions,
hallucinations, and incoherent speech were added for a
total positive symptom score (maximum 12), and scores
for poverty of speech, flat affect, and psychomotor
retardation were added for a total negative symptom
score (maximum 12).

Length of illness, defined as current age minus
age at onset, and months spent on a psychiatric
inpatient ward were also determined.

Pearson product-moment was used for correlation
analysis. Two-way ANOVA with diagnosis and sex factors
was also used. All analyses yielding a p value of .05
or less were considered statistically significant.

RESULTS

Table 1 summarizes the descriptive data concerning
morphological measurements when all subjects were
pooled and when they were stratified by sex.

Table 1. Cortical Area (CA), Ventricular Brain Ratio
(VBR) and Corpus Callosum Brain Ratio (CCBR) Measured
on Midsagittal and Axial Cut.

Area and Sex	Schizophrenics		Controls	
	N	Mean±SD	N	Mean±SD
CA				
All subjects	30	84.89±8.56	25	86.15±9.02
Males	23	86.60±8.73	18	86.55±9.98
Females	7	79.27±5.12	7	85.11±6.43
VBR				
All subjects	30	5.23±2.23	25	3.80±1.12
Males	23	5.27±2.33	18	3.73± .79
Females	7	5.08±2.01	7	4.00±1.78
CCBR				
All subjects	30	6.85±1.22	25	7.51±1.08
Males	23	6.78±1.32	18	7.63± .94
Females	7	7.08± .83	7	7.20±1.42

Table 2 examines these data statistically through analysis of variance.

Table 2. Analysis of Variance of Diagnosis and Sex in Cortical Area (CA), VBR and CCBR (30 schizophrenics patients vs. 25 healthy volunteers).

Area	Source	F	P
CA	Diagnosis	.394	.533
	Sex	2.821	.099
	Diagnosis by sex	1.219	.275
VBR	Diagnosis	8.067	.006
	Sex	.003	.959
	Diagnosis by sex	.169	.683
CCBR	Diagnosis	4.269	.044
	Sex	.022	.884
	Diagnosis by sex	1.010	.320

For the cortical area there was a trend toward a sex effect and indeed female schizophrenics showed a smaller CA. For VBR there was highly significant diagnostic effect as well as for CCBR.

We are, of course, interested in determining whether the possible brain abnormalities had any relationship with the clinical picture.

No statistically significant correlation was seen between negative-positive symptom evaluation and any of the cerebral measurements.

When patients were divided into two groups defined as those with VBR exceeding mean VBR+2SD of the control group and those with VBR inside this threshold, no differences in clinical symptomatology emerged between these two groups. No other significant relationships were observed between brain morphological variables and length of illness or cumulative hospitalization.

DISCUSSION

The VBR findings confirm those of previous CT studies [2].

Our findings of smaller CCBR in schizophrenics as compared to healthy controls are relevant for discussion. The main findings about CC structure in schizophrenia come from post-mortem studies and, only recently, from MR studies. In replicating the original post-mortem report of increased callosal thickness in schizophrenia [3], Bigelow et al.[4] reported that 21 early onset schizophrenic brains showed a significantly greater callosal thickness as compared to late onset schizophrenia,neurological diagnoses and other psychiatric diagnoses. They hypothesized that a pathological process (i.e. sub-clinical viral infection)may lead to a slightly increased tissue volume and later,after cell loss,to a relatively normal dimension. Both Rosenthal [3] and Bigelow [4] measured the thickness of the CC in coronal slices 0.5 cm lateral to the midline, and Nasrallah et al. [9] reported an increased callosal thickness in female schizophrenics measured by MR on the sagittal plane. Mathew et al. [16] in a MR study showed that schizophrenics had significantly longer CC than controls.

We used the CCBR because this ratio can control the effect of brain size in clinical samples and it is a more reliable indicator of the quantitative relationship of callosal area per brain tissue area unit.

We report a smaller CCBR on the midsagittal plane and a larger VBR on the axial plane in the patient group as compared to controls. Nasrallah et al. [9] did not confirm our CCBR finding, reporting a higher CCBR in schizophrenics as a whole, but not in the male patients, and an increase of CC anterior and mid-width without CCBR increase in females. Our mid-sagittal measures are very close to those reported by Nasrallah et al. [9] and Andreasen et al. [12], suggesting a high methodological agreement between the two measurement techniques.

The discrepancy in CC findings between this paper and the previous ones does not seem to derive from measurement differences and sampling factors are also thought to have some influence on this result.An alternative explanation of our findings could be taken from developmental neurobiology. If the Lateral Ventricular Enlargement (LVE) we report does reflect brain tissue reduction, the smaller callosal dimension we report may be due to an acceleration of the physiological regressive elimination of axon collaterals occuring early in fetal or newborn stages [17], associated with a damaging event in the structures or system involved in schizophrenia. From this perspective one may hypothesize that abnormalities in the CC area may be an adaptive morphofunctional reaction to a pathological process appreciated as LVE, eliciting a subtle disruption of the brain's internal morphology.

From this perspective the naturally-occuring regressive events [17] may be modified, leading to a "reshaping" of the callosal dimension.

A possible shortcoming of gross morphological studies such as this is that measurements of cross-sectional areas are regarded as studies of fibre number and nerve fibres themselves are assumed to be invariable in structure [18].

Ultrastructural information on the relative numbers and density of myelinated and unmyelinated fibres of Corpus Callosum are needed before we can interpret further our findings.

Further studies may add clues to the possible existence of a structural callosal pathology in schizophrenia or even whether the failure in inter-hemispheric transfer reported in schizophrenia [8] may be correlated to the putative abnormal callosal structure.

Later MR studies, allowing brain visualization in multiple planes, may add further advances in understanding whether VBR and CC abnormalities represent two different dimensions of brain pathology in schizophrenia.

REFERENCES

1. Haug, JO (1982). Pneumoencephalographic evidence of brain atrophy in acute and chronic schizophrenic patients. Acta psychiat scand, 66, 374

2. Shelton, RC and Weinberger, DR (1986). X-ray computerized tomography studies in schizophrenia: a review and synthesis. In: Nasrallah, HA and Weinberger, DR (eds.) "The Neurology of Schizophrenia". p.207. (Amsterdam: Elsevier)

3. Rosenthal, R and Bigelow, LB (1972). Quantitative Brain Measurements in Chronic Schizophrenia. Brit J Psychiat 121, 259

4. Bigelow, LB, Nasrallah, HA and Rausher, FP (1983). Corpus Callosum Thickness in Chronic Schizophrenia. Brit J Psychiat 142, 284

5. Green, P (1978). Defective interhemispheric transfer in schizophrenia. J Abnorm Psychol 87, 472

6. Beaumont, J and Dimond, S (1973). Brain disconnection and schizophrenia. Br J Psychiatry 123, 661

7. Carr, SA (1980). Interhemisperic transfer of stereognosic information in chronic schizophrenia. Br J Psychiatry 36, 53

8. David, AS (1987). Tachistoscopic tests of colour naming and matching in schizophrenia: evidence for posterior callosum dysfunction? Psychol Med 17, 621

9. Nasrallah, HA, Andreasen, NC, Coffman, JA et al (1986). A Controlled Magnetic Resonance Imaging Study of Corpus Callosum Thickness in Schizophrenia. Biol Psychiatry 21, 274

10. Smith, RC, Baumgartner, R and Calderon M (1987). Magnetic Resonance Imaging Studies of the Brains of Schizophrenic Patients. Psychiatr Res 20, 33

11. Rossi, A, Stratta, P, Gallucci, M, Passariello, R, Casacchia, M (1987). Corpus Callosum in Schizophrenia. Biol Psychiatry 22, 1043

12. Andreasen, NC, Nasrallah, HA, Dunn, V et al (1986). Structural Abnormalities in the Frontal System in Schizophrenia. A Magnetic Resonance Imaging Study. Arch Gen Psychiatry 43, 136

13. Nasrallah, HA, Olson, SC, Coffman, SB et al (1988 in press).Magnetic Resonance Brain Imaging, Perinatal Injury and Negative Symptoms in Schizophrenia. Schizophr Res

14. Oldfield, RC (1971). The assessment and analysis of handedness: the Edinburgh Inventory. Neuropsychologia 9, 97

15. Krawiecka, M, Goldberg, D and Vaughan, M (1977). A standardized psychiatric assessment scale for rating chronic psychotic patients. Acta psychiat scand 55, 299

16. Mathew, RJ, Partain, CL, Prakash, R et al (1985). A study of the septum pellucidum and corpus callosum in schizophrenia with MR imaging. Acta psychiatr scand 72, 414

17. Cowan, WM, Fawcett, JW, O'Leary, DDM et al (1984). Regressive Events in Neurogenesis. Science 255, 1258

18. Demeter, S, Ringo, JL and Doty, RW (1988). Morfhometric analysis of the human corpus callosum and anterior commissure. Human Neurobiol 6, 219

3
Brain density patterns in schizophrenia

A. Vita, C. Sina, E. Sacchetti, G. Valvassori, A. Calzeroni, G.M. Giobbio, G. Invernizzi, V. Bernasconi and C.L. Cazzullo

INTRODUCTION

In the last decade there have been many studies on the morphological cerebral characteristics of schizophrenic patients, particularly with computerized tomography (CT) (1,2). The most frequently studied parameters have been cerebral ventricular size and cortical atrophy, both of which have been shown to be fairly different from the norm (1-4).

Few studies, however, have examined the density of the cerebral parenchyma. This has partially been due to the technical requirements necessary for establishing cerebral X-rays absorbtion density.

Further, the studies that have been undertaken have had notably discordant results. In the first study on this subject, Golden et al. (5) reported the presence of a diffuse brain hypodensity in schizophrenic patients, that was more pronunced in the left frontal region.

Measuring the density of circumscribed circular areas in the white and gray matter, Largen et al. (6,7) noted a greater radiodensity of the right hemisphere particularly as regarded the white matter, but his subjects did not significantly differ from controls.

Coffman and Nasrallah (8) found a global reduced density in the most cranial of the three sections studied.

Dewan et al. (9) observed a significant hyperdensity of the periventricular gray nuclei they studied. In a subsequent study (10) the same investigators also noted a non significant increase of the parenchymal density of cortical sites in the schizophrenic patients they examined.

In a recent study conducted by use of a sophisticated method of automatic reading of the tissue density able to exclude the liquor spaces from the analysis, Reveley et al. (11) found a hemispheric asymmetry of the cerebral density in the patients (schizophrenic members of pairs of monozygote twins) with relative hypodensity of the left hemisphere; however, no substantial absolute differences from controls were noted.

It is difficult to interpret these results in a uniform way, and the field of densitometric studies still appears to be rather confused.

Further, the ever-expanding knowledge of the many artifacts that affect the computer-derived numerical values of absorbtion density (12,13) has demonstrated the technical and methodological limits of many studies, particularly those undertaken with first generation tomography.

All of these means that it is necessary to delve further into this field with studies that use techniques more sensitive to controlling the artifacts connected with the techniques themselves and have "tighter" designs.

This report presents the preliminary results of an ongoing study on the density of specific cerebral areas and nuclei in the course of schizophrenia.

SUBJECTS AND METHODS

Subjects

The study was conducted with 21 patients, 13 males and 8 females, average age 28.7+6.3 years. The criteria for including patients were: a)diagnosis of schizophrenic disorder as defined by DSM-III; b)absence of medical or neurological illness; c)no history of alcoholism, drug abuse, cranial trauma with loss of consciousness, epilepsy or ECT; d)no corticosteroid treatment in the 3 months prior to the study.

Fourteen subjects (8 males, 6 females, mean age 29.8+6.9 years) were chosen as controls. Of these, 8 were healthy volunteers recruited from among hospital students and staff and 6 were subjects who had undergone a cerebral CT following a minor head trauma (without loss of consciousness) as a result of an accidental fall or car accident.

The criteria for inclusion in the control group were: a)absence of signs and symptoms of cerebral nervous system lesions at the time of the study; b)no past or present major psychiatric disorders; c)criteria b) to d) as described for the patients.

Methods

The tomographic studies were performed without a contrast medium on a GE 9000 series II tomograph. Twelve cerebral tomographic sections were obtained for each patient and control. These were parallel to the canthomeathal line and each was 10 mm thick.

Parenchimal density. The density of specific "regions of interest" (ROI) were measured. These were standardized with respect to shape, size, and anatomical reference points singled out in different cerebral areas and circumscribed on the screen of the tomographic computer by means of a suitable cursor. The following anatomical structures were selected for analysis: a)the head of the caudate nucleus; b)thalamus; c)periventricular substance and d)white hemispheric substance.

All the measurements were taken on three standard tomographic sections:
- section A - passing through the frontal horns of the lateral ventricles and the third ventricle;
- section B - passing through the body of the lateral ventricles;
- section C - passing through the centrum semiovale (usually the second or third section above B, proceeding in a craniocaudal direction).

Specifically:
a) with relation to section A the following structures were evaluated:
- head of the caudate nucleus: average density of a circular area of 13.5 mm^2, 0.5 cm lateral to the most concave edge of the anterior horns of lateral ventricles (bilaterally);
- anterior thalamus: density of a circular area of 13.5 mm^2, 0.5 cm lateral to the edge of the third ventricle at its median point (bilaterally);
- posterior thalamus: area 1 cm behind the preceding one along the major diagonal of the thalamic nucleus (bilaterally);
b) on section B, we evaluated the white periventricular substance: density of a circular area of 13.5 mm^2, 1.2 cm lateral to the external edge of the body of the lateral ventricles at its median point (bilaterally);
c) evaluation on section C were as follows:
- white substance of the centrum semiovale, anterior, middle and posterior: density of a circular area of 44 mm^2 placed in the intermediate zone of the centrum semiovale at the third anterior, middle and third posterior points of the cerebral sagittal diameter.

Once these regions were circumscribed the density was automatically furnished by the tomograph as the mean ($\pm$ standard deviation) of the attenuation values (Hounsfield units) of the volume units (voxel) included in that area.

Numerous expedients were used to reduce the many artifacts that can affect the densitometric evaluation of CT images (12,13):

a- there were no changes in the software or tomographic parts during the course of the study;
b- the calibration of the instrument was performed monthly using
an identical procedure carried out by the same team;
c- the study was completed in just a few months and both patients
and controls were examined in substantially similar time periods;
d- images showing any visible artifact (by movement or not) were
excluded from the analysis.

RESULTS

Table 1 shows the results of the analysis of radiodensity of the
various cerebral areas examined.
    We found a general increase of tissue density values in schi-
zophrenic subjects. This reached statistical significance for the
anterior thalamus, particularly in the right hemisphere (37.51+4.54
vs 34.74+1.65 H.U.; F=4.76, d.f.: 1,33; p=.03) but also in the
left one (36.75+4.46 vs 34.39+1.69 H.U.; F=3.53, d.f.:1,33; p=.06).
    Other sites affected by this density increase were the poste-
rior thalamic region and the periventricular substance even though
the differences as compared with controls were not statistically
significant.
    The further analysis of the data showed a notable lack of homo-
geneity of the variance of the measures obtained from the patients
as contrasted with the controls. Thus, the variability of the density measures was higher in the patients than in the normal controls.
    The degree of correlation between the density of the different
cerebral areas examined was rather high in the controls (intraclass correlation coefficient: r=.42; F=11.0, d.f.: 13,182; p<.001)
but almost nil in the patients (intraclass correlation coefficient:
r=.016; F=1.33, d.f.: 13,280; p=NS).
    Consequently, we undertook a study to ascertain the factors
possibly responsible for this variability of the results. The variables taken into consideration were demographic (patient age
and sex), clinical (age at onset, duration of illness, family history of schizophrenia in first degree relatives, intellectual level as measured by the Wechsler Adult Intelligence Scale, prevalent
symptomatology as measured by the BPRS (14) and the Scales for Assessment of Positive and Negative Symptoms (15)), and other neuromorphological variables (VBR, cortical atrophy). Two other variables which have been shown to influence cerebral density values
also were taken into account: head size and the time at which the
examinations were administered. The first of these variables conditions the extention of the areas affected by the phenomenon of

"hardening" of the photonic beam, more marked in areas adjacent to the cranial theca. The second involves differences in absorb—tion density measures related to calibration errors of the instruments (13). None of the variables examined, however, correlated significantly with the radiodensity values of the specific cerebral areas studied.

Table 1. Brain density measures in schizophrenic patients and in healthy controls.

AREA	SIDE	CONTROLS mean*+ SD	SCHIZOPHRENICS mean*+ SD	F (1,33)	p
CAUDATE	L	37.38+2.50	36.04+4.92	.72	NS
	R	37.68+1.79	38.43+5.16	.27	NS
THALAMUS(ant)	L	34.39+1.69	36.75+4.46	3.53	.06
	R	34.74+1.65	37.51+4.54	4.76	.03
THALAMUS(post)	L	34.38+1.69	36.40+4.46	1.91	NS
	R	34.81+2.37	37.06+5.00	2.83	NS
PERIVENTRICULAR AREA	L	33.15+1.82	35.06+4.12	2.64	NS
	R	33.35+1.82	34.96+4.18	1.84	NS
CENTRUM SEMIOVALE (ant)	L	34.66+1.79	35.99+6.06	.64	NS
	R	34.30+1.63	35.88+5.71	1.01	NS
CENTRUM SEMIOVALE (mid)	L	32.51+1.61	34.24+5.69	.004	NS
	R	32.45+1.56	34.32+5.56	1.50	NS
CENTRUM SEMIOVALE (post)	L	32.51+1.66	34.53+5.40	1.82	NS
	R	32.75+1.60	34.54+6.18	1.11	NS

* Hounsfield Units

Subsequently, we compared the radiodensity of the two cerebral hemispheres in patients and controls. These results are summarized in Table 2.

While controls did not show differences in the density of the two hemispheres for any of the areas studied, we observed a significant difference in the hemispheric densities of the caudate nucleus and anterior thalamus among the patients.

In both cases there was a hypodensity of the left hemisphere as compared with the right one. This was particularly pronounced relative to the caudate nucleus (t=5.49; p<.001).

Table 2. Interhemisphere comparison of brain tissue density measures in schizophrenic patients and in healthy controls.

AREA	GROUP°	LEFT HEMISPHERE mean*+SD	RIGHT HEMISPHERE mean*+SD	t§ (paired)	p
CAUDATE	C	37.38+2.50	37.68+1.79	.58	NS
	S	36.04+4.92	38.43+5.16	5.49	.001
THALAMUS(ant)	C	34.39+1.69	34.74+1.65	.89	NS
	S	36.75+4.46	37.51+4.54	2.48	.021
THALAMUS(post)	C	34.38+1.69	34.81+2.37	.65	NS
	S	36.40+4.46	37.06+5.00	1.05	NS
PERIVENTRICULAR AREA	C	33.15+1.82	33.35+1.82	.09	NS
	S	35.06+4.12	34.96+4.18	.37	NS
CENTRUM SEMIOVALE (ant)	C	34.66+1.79	34.30+1.63	.07	NS
	S	35.99+6.06	35.88+5.71	.27	NS
CENTRUM SEMIOVALE (mid)	C	32.51+1.61	32.45+1.56	.50	NS
	S	34.24+5.69	34.32+5.56	.47	NS
CENTRUM SEMIOVALE (post)	C	32.51+1.66	32.75+1.60	.84	NS
	S	34.53+5.40	34.54+6.18	.03	NS

* Hounsfield Units
§d.f.= 13 for controls and 20 for schizophrenic patients
o C = controls   S = schizophrenics

DISCUSSION

The rather small number of cases studied is the most important limitation of the study and makes it impossible to generalize from
its results.

Further, the tomographic measures of density are subject to
many artifacts inherent in the method itself and capable of alte-
ring the values obtained (12,13). This study, however, took several
steps towards reducing these artifacts.

Thus, the visible artifacts, by movement or not, were avoided
by excluding the slices showing such changes.

In addition, the operative conditions of the tomographic system
were rigorously mantained constant.

Moreover, there was no relationship between the density values
and the time of execution of the scan during the course of the study and this confirms the slight influence of calibration and grada-
tion factors on the results obtained.

Again, this study avoided density measurements of areas adia-
cent to the cranial theca and of apical regions, areas greatly affected by the so-called "beam hardening effect" (13). For the same
reasons, the head size, also capable of influencing the cerebral
density estimate, was checked as possible source of variation of
the measures obtained.

Last, "partial volume effect", due to the coexistence of different tissues (white, gray matter, cerebrospinal fluid) in the
area subjected to scanning, was kept under control by avoiding measurement of cortical areas (near the subarachnoid spaces) and areas
immediately near the ventricular edges. Even if this effect was
not always eliminated, particularly in the case of measuring the
periventricular nuclei, it is reasonable to think that it only randomly affected the subjects studied, independently of diagnosis,
and thus does not explain the differences found between the diagnostic groups.

Keeping the limitations described in mind, it seems important
to comment on the three fundamental densitometric findings we obtained in schizophrenic patients: a hyperdensity of the thalamus,
a variability of densitometric values rather higher than the norm
and an asymmetry of the density of some periventricular nuclei,
particularly the caudate, absent in the normal subjects.

Finding a hyperdensity of the anterior thalamic regions is
in accordance with the only tomodensitometric study that has evaluated this nucleus (9), with a method similar to our one.

The observed hyperdensity affected the entire thalamic nucleus, but appeared more pronounced in its anterior sites, that
is, precisely in the regions surrounding the third ventricle.

Finding a change in the thalamic radiodensity is compatible both

with Bowman and Lewis' (16) observations that various types of thalamic pathology can be associated with clinical symptoms and signs
typical of schizophrenia, and with the numerous demonstrations of
an involvement of this nucleus in the neuropathology of schizophrenia (17,18).

The interpretation of this result appears complex.

It seems possible to hypothesize, with Dewan et al. (9), that
the hyperdensity may reflect a changed chemical and/or cellular
composition of the nervous tissue.

Contrarily, it does not seem probable that the hyperdensity
was due only to the effect of the enlargement of the ventricular
system with consequent "crushing" and thus thickening of the surrounding areas since we and others (19) did not observe any relationship between ventricular dimensions and parenchymal radiodensity.

If an underlying histopathologic change might represent the
most probable substrate of the hyperdensity of the thalamic regions, the nature of such a change remains completely hypothetical,
particularly because of the present lack of precise knowledge of
the anatomopathologic correlations of the tomodensitometric findings.

A possible interpretation is provided by recent neuropathologic results obtained by Stevens (17). These indicate the presence
of fibrillary gliosis in schizophrenic patients, irregularly distributed in different cerebral areas, but particularly in the periventricular regions. According to Stevens, the thalamic nuclei also
would be involved in this process of astroglial substitution of
the cerebral tissue.

The presence of gliosis, already demonstrated by several investigators in various cerebral regions in the course of schizophrenia (18), obviously is indicative of a previous neuropathological process, probably of an inflammatory nature (17) and gliosis
signals its location.

The demonstration of a process of periventricular gliosis in
post-mortem sudies therefore offers direct support for the hypothesis that in some cases schizophrenic disturbances can be the result
of a pathologic cerebral process primed by neurologically harmful
environmental factors.

The lack of comparative studies of in vivo tomographic findings and post-mortem observations, however, impedes conclusions
even if at least one study (20) showed that a condition of tomographic hyperdensity (evaluated subjectively) was correlated with a
gliotic process which was confirmed by autopsy.

Alongside the finding of hyperdensity of the thalamus, the
present study demonstrated a higher variability of density measures
in patients than in controls. For one thing, this suggests that

hyperdense parenchymal zones also are found outside of the thalamus and affect the majority of the structures studied. It further suggests that areas of tissue hypodensity also are found in these same structures and thus both extreme densitometric irregularities in individual patients and a notable interindividual variability of the tomographic picture are encountered. The latter finding substracts specificity and certainly any pathognomonic significance to the tomodensitometric findings in the course of schizophrenia but confirms structural brain changes substratum in this disease.

Finally, the observation of an asymmetry of density of the caudate with relative hypodensity of the left side was an unexpected result which, however, confirms and better localizes previous observations of an hypodensity relative to the left hemisphere, as compared to the right, in schizophrenic patients (5,6,7,11).

This finding also is compatible with the behavioral studies indicating specific lateralization defects and, more often, a dysfunction of the left hemisphere in schizophrenic patients. These functional asymmetries can, moreover, give some contribution to the interpretation of our results.

The evidence for a hemispheric dysfunction (left) in the course of schizophrenia derives from a host of studies - clinical (21), electrophysiological (EEG) (22), neurochemical (23), neuromorphological (6,7,11) - and from studies on the cerebral metabolism and blood flow (24,26).

Our observation of an asymmetry of caudate density and, to a lesser extent, of thalamus coherently fits into this context.

A very theoretical but worthy hypothesis is that the hypodensity of the caudate nucleus is to be attributed to, or is accompanied by, its distortion or morphologic modification or to its dislocation. In this case, the underlying structural aberration could be a drawback to a developmental defect early in the individual's life in accordance with a "neurodevelopmental" theory to which many investigators now suscribe (25,27). This theory would hold that the lateralization defect is a crucial element in the "dis-developmental" process and would be responsible for the future development of schizophrenic pathology.

It is clear from these considerations the notable heuristic potential inherent in the study of the radiodensity of specific brain nuclei and areas. It is to this potential that we should entrust the verification of many of the present hypotheses on the etiopathogenesis of schizophrenia.

REFERENCES

1.   Goetz, KL and van Kammen, DP (1986). Computerized axial tomography scans and subtypes of schizophrenia. J.Nerv.Ment.Dis., 174, 31-41

2.   Shelton, RC and Weinberger, DR (1986). X-ray computerized tomography studies in schizophrenia: a review and synthesis. In: Nasrallah, HA and Weinberger, DR (eds.)"The Neurology of Schizophrenia". pp. 207-250. (Amsterdam: Elsevier Science Publ.)

3.   Sacchetti, E, Vita, A, Calzeroni, A, et al. (1987). Neuromorphological correlates of schizophrenic disorders: focus on cerebral ventricular enlargement. In: Cazzullo, CL, Invernizzi, G, Sacchetti, E, and Vita, A, (eds.) "Etiopathogenetic Hypotheses of Schizophrenia". pp. 67-93. (Lancaster: MTP Press)

4.   Vita, A, Sacchetti, E, Calzeroni, A et al. (1988). Cortical atrophy in schizophrenia. Prevalence and associated features. Schizophrenia Res., 1, 329-337

5.   Golden, CJ, Graber, B, Coffman, J, et al. (1981). Structural deficits in schizophrenia, identification by computed tomographic scan density measurements. Arch. Gen. Psychiatry, 3, 33-39

6.   Largen, JW, Calderon, M and Smith, RC (1983). Asymmetries in the density of white and gray matter in the brains of schizophrenic patients. Am. J. Psychiatry, 140, 1060-1062

7.   Largen, JW, Smith, RC, Calderon, M, et al. (1984). Abnormalities of brain structure and density in schizophrenia. Biol. Psychiatry, 19, 991-1013

8.   Coffman, JA and Nasrallah, HA (1984). Brain density patterns in schizophrenia and mania. J. Affect. Disord., 6, 307-315

9.   Dewan, MJ, Pandurangi, AD, Lee, SH, et al. (1983). Central brain morphology in chronic schizophrenic patients: a controlled CT study. Biol. Psychiatry, 18, 1133-1140

10.   Dewan, MJ, Pandurangi, AH, Lee, SH, et al. (1986). A comprehensive study of chronic schizophrenic patients: I.Quantitative computed tomography. Acta Psychiatr. Scand., 73, 152-160

11.   Reveley, MA, Reveley, AM and Baldy, R (1987). Left cerebral hemisphere hypodensity in discordant schizophrenic twins: a controlled study. Arch. Gen. Psychiatry, 44, 625-632

12.   Jacobson, RR, Turner, SW, Baldy, RE, et al. (1985). Densitometric analysis of scans. Important sources of artifact. Psychol. Med., 15, 879-889

13.   Mc Cullough, EC (1977). Factors affecting the use of quantitative information from CT scanner. Radiology, 124, 99-105

14.   Overall, JE and Gorham, DR (1962). The brief psychiatric rating scale. Psychol. Rep., 10, 799-810

15.   Andreasen, NC (1987). Moscarelli, M and Maffei, C (eds.) "Schizofrenia: scale per la valutazione dei sintomi positivi e

negativi". Italian Edition. (Milano: Libreria Cortina)
16. Bowman, M and Lewis, MS (1980). Sites of subcortical damage in diseases which resemble schizophrenia. Neuropsychologia, 18, 597-601
17. Stevens, JR (1982). Neuropathology of schizophrenia. Arch. Gen. Psychiatry, 39, 1131-1139
18. Zec, RF and Weinberger, DR (1986). Brain areas implicated in schizophrenia: a selective review. In: Nasrallah, HA and Weinberger, DR (eds.) "The Neurology of Schizophrenia". pp. 175-206 (Amsterdam: Elsevier Science Publ..)
19. Jernigan, TL, Zatz, LM, Moses, JA, et al. (1982). Computed tomography in schizophrenics and normal volunteers. I: Fluid volume. Arch. Gen. Psychiatry, 39, 765-770
20. Weisberg, L (1981). Computed tomography findings in intracranial gliosis. Neuroradiology, 21, 253-257
21. Flor-Henry, P (1976). Lateralized temporal-limbic dysfunction and psychopathology. Ann. N. Y. Acad. Sci., 280, 777-795
22. Gruzelier, J (1979). Synthesis and critical review of the evidence for hemispheric asymmetries of functions in psychopathology. In: Gruzelier, J and Flor-Henry, P (eds.) "Hemispheric Asymmetries of Function in Psychopathology". pp. 193-214. (Amsterdam: Elsevier)
23. Reynolds, J (1983). Increased concentrations and lateral asymmetry of amygdala dopamine in schizophrenia. Nature, 305, 527-529
24. Ingvar, DM and Franzen, G (1974). Abnormalities of cerebral blood flow distribution in patients with chronic schizophrenia. Acta Psychiatr. Scand., 50, 425-462
25. Weinberger, DR (1987). Implications of normal brain development for the pathogenesis of schizophrenia. Arch. Gen. Psychiatry, 44, 660-669
26. Weinberger, DR, Berman, KF and Zec, RF (1986). Physiological dysfunction of dorsolateral prefrontal cortex in schizophrenia: I. Regional cerebral blood flow (rCBF) evidence. Arch. Gen. Psychiatry, 43, 114-125
27. Crow, TJ (1986). Left brain, retrotransposons, and schizophrenia. Br. Med. J., 293, 3-4

ACKNOWLEDGMENT

This research was supported in part by C.N.R. grant no. 87.00229.56

4
Lateral ventricular size and clinical features in schizophrenia

C. Maggini, M. Guazzelli, S. Starnini, M. Ardito, L. Rosi, F. Busoni and E. Camerini

INTRODUCTION

Many researchers have concluded that a subset of schizophrenic patients has brain ventricle enlargement at the Computerized Tomographic (CT) examination [1-15] and that such brain structural abnormality is correlated with prevalence of negative symptoms [16,17], response to antipsychotic drugs [11,18], severe long-term outcome [17], and high impairment of neuropsychological performances [1,3,19]. Several controlled CT studies however did not confirm these findings [20-23]. Moreover, it has not been ruled out that brain ventricle abnormalities might be related rather to external variables such as institutionalization or previous somatic treatments, than to the pathophysiologic mechanisms underlying this psychiatric disorder.

Therefore relationships between clinical symptomatology, neuropsychological performances, long-term course of schizophrenia and cerebral ventricular size still need to be clarified.

In the present paper we report VBR and baseline clinical data of the first 30 patients included in an on-going long-term prospective study aimed to investigate CT brain structure and its relationships with the DSM III schizophrenic types, positive and negative symptoms, neuropsychological performances, treatment outcome and clinical course of the illness on a large sample of non institutionalized schizophrenics.

MATERIAL AND METHOD

Schizophrenic patients were recruited among those consecutively admitted to the Male Unit of the Clinica Psichiatrica of the University of Pisa who consented to participate to the study and met the following criteria: DSM-III [24] diagnosis of Schizophrenic Disorder, age between 20 and 40 years, duration of illness between 2 and 6 years, no previous psychiatric hospitalization lasting more than two months and negative history of brain trauma, mental retardation, neurologic and systemic disorders, drug and alcohol abuse, electroconvulsive or insulin therapy, and corticosteroid

33

treatment.   Structured   Clinical   Interview   for   DSM-III [25]
administered   by   two   senior   psychiatrists   was   used   for   the
diagnosis.   The Scale of Assessment of Negative Symptoms of
Andreasen   [26]   (the   Attention   was   evaluated   in   the
neuropsychological investigation; in the clinical investigation the
subscale   of   Attention   was   not   employed)   and   the   Scale of
Assessment of Positive Symptoms [27] were administered within the
first 10 days following hospitalization and criteria of Andreasen
[28] for positive, negative and mixed schizophrenic subgroups were
employed.   Demographic   and   clinical   data,   consisting of age,
duration of illness, educational level reached at the onset of the
illness, number of hospitalizations, classification according to
the DSM-III schizophrenic types and to the Andreasen criteria are
reported in table 1 and 2.

Table 1.   Demographic and clinical data (Part A)

	Total sample	Disorganized	Undifferentiated	Paranoid	Residual
N	30	8	7	6	9
Age	25±3.4	22.5±2.4	26±2.4	25.8±3.8	25.9±3.8
Duration of illness	4.1±1.5	2.6±1.3	4.2±1.4	5.0±1.1	4.6±1.3
High education	12	5	2	3	2
Low education	18	3	5	3	7
1st or 2nd hospitalization	18	6	6	2	4
Multiple hospitalization	12	2	1	4	5

The age of the patients ranged between 20 and 31 years, mean
age was 25 ± 3.4 years, educational level was low (i.e. 8 years
mandatory education according to Italian Educational System) in 18
patients and high (High School or University) in 12 patients, 18 of
them were at their first or second hospitalization, and 12 had
multiple hospitalizations.   Average duration of illness was 4.1 ±
1.5 years. On the basis of the clinical assessment, 8 patients were
classified as disorganized, 7 as undifferentiated, 6 as paranoid
and 9 as residual according to DSM-III.
According to Andreasen criteria 8 patients were positive, 12
negative   and   10   mixed.   Schizophrenic   subgroups   showed   no

difference between them for any of the reported parameters.

Table 2.   Demographic and clinical data (Part B)

	Total sample	Positive	Negative	Mixed
N	30	8	12	10
Age	25±3.4	25.2±2.5	24.2±4.0	25.7±3.3
Duration of illness	4.1±1.5	4.4±1.4	3.6±1.5	4.2±1.7
High education	12	4	4	4
Low education	18	4	8	6
1st or 2nd hospitalization	18	5	7	6
Multiple hospedalization	12	3	5	4

After clinical   evaluation, each patient underwent brain CT scan investigation at the Clinica Radiologica of the University of Pisa; radiologic   researchers   were blind   to   the clinical characteristics of the patients.   CT Scans were performed without contrast material on a General Electric (G.E.) CT Max third generation with 320x320 matrix. Sections were made at 10 mm intervals.   The ratio between cerebral   area (automatically calculated after manual tracing of the brain perimeter on the monitor)   and the lateral   ventricular   areas (automatically calculated after manual tracing of the ventricles perimeter on the monitor) in the slice where the ventricles had the largest size was defined Ventricular Brain Ratio (VBR).

Brain CT scans of 30 male subjects, selected according to age from the records of healthy subjects of the Clinica Radiologica of the University of Pisa, were used as control data.   At the inclusion of each schizophrenic patient a control subject with age as close as possible to that of the patient (within 2 years), was picked out of the records and matched with the patient.   Mean age of the 30 control subjects so far collected was 25.5 ± 3.8 years, and it was comparable to that of the schizophrenics.   As generally accepted in literature [2,4,5], the cut off between normal and abnormal VBR of the schizophrenics was the VBR mean value plus 2 SD of the control group.

Goodness-of-fit test, two-tailed Student's t-test and linear correlation analysis of the Statistical Analysis System – S.A.S. [29] were used for statistical evaluation.

RESULTS

In the control group VBR ranged between 1.9 and 8.5 with a mean of 4.7 $\pm$ 1.9 ; it was normally distributed (chi-square for goodness-of-fit to normal distribution: 1.7, p=NS) and positively correlated with age, although not at statistical level (r=0.25). According to the definition lateral ventricles were abnormally enlarged when VBR was higher than 8.5 (cut off value).

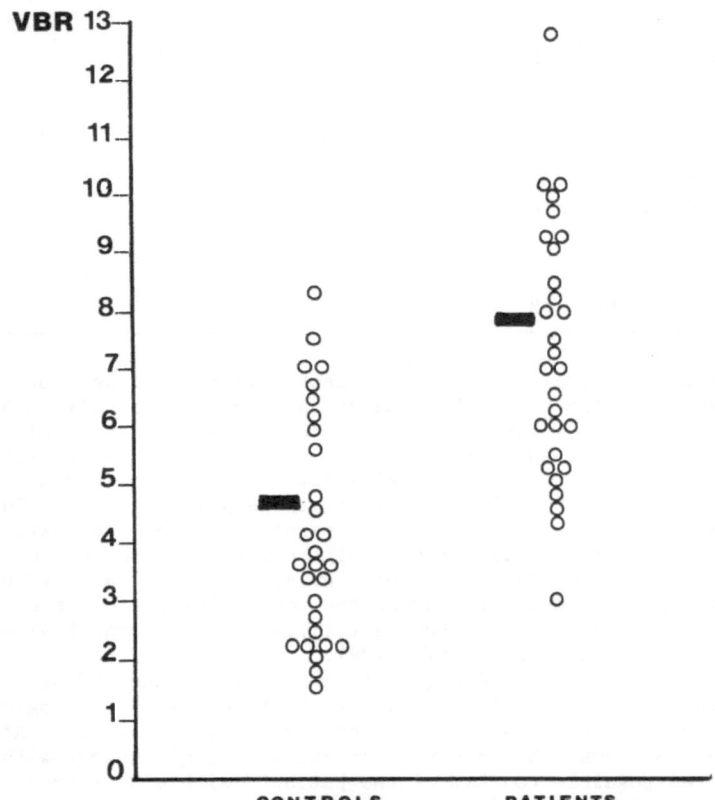

FIGURE 1   Distribution of VBR of the 29 schizophrenics and of the 29 age matched controls subjects:
    ▄▄▄ = VBR mean value.
    Differences between the two groups were statistically significant: t=4.9, p<.001 at two-tailed Student's t-Test.

In the schizophrenic group VBR ranged between 3.0 and 19.8 with a mean value of 7.8 ± 3.2. VBR of 19.8 was found in a 22 year old, low educated patient; he has been ill in the last 2 years and he was actually classified as residual according to the DSM-III and as negative according to the Andreasen criteria. Even after a careful control of the patient's clinical data and an accurate interview with its family we could not reach any evidence of actual or previous somatic diseases or substance abuse which could correlate with CT data. However we arbitrarily decided not to include this patient in the analyses. Thus VBR of the 29 schizophrenics ranged between 3.0 and 12.9, with an average of 7.3 ± 2.2 and it was normally distributed (chi-square for goodness-of-fit to normal distribution: 1.7, p=NS). In this sample VBR was neither significantly correlated with age (r= -0.13) nor with duration of illness (r=-0.19).

In 9 patients (31% of the whole sample) VBR was higher than cut off value for abnormal enlargement.

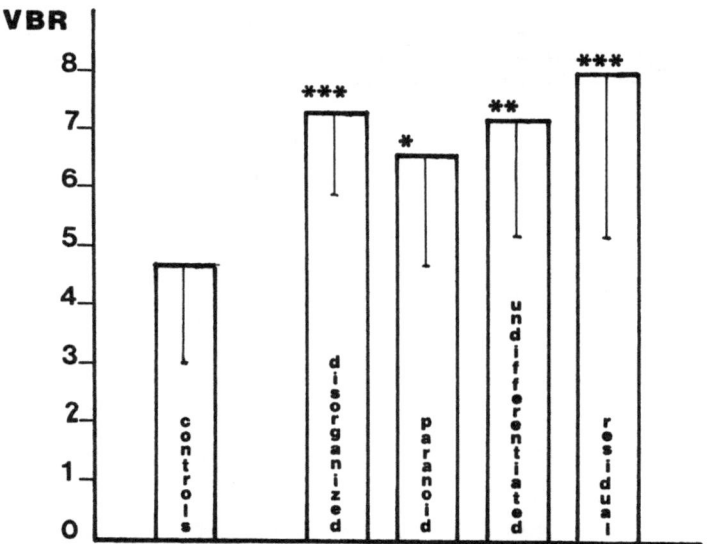

FIGURE 2   VBR mean values and SD of the 29 control subjects and the 29 schizophrenic patients grouped according to DSM-III schizophrenic types.
Results of two-tailed Student's t-Test between groups:
* = p<.05, ** = p<.005 and *** = p<.001 vs control group
No significant differences between the patient groups.

As shown in Fig. 1, in the schizophrenic group, compared with the control group, VBR was significantly higher (t=4.9, p<.001). VBR was significantly higher in the schizophrenics even when the 9 patients with abnormally high values were excluded, and the 20 patients with VBR under the cut off value were compared both with the whole control sample (6.2 $\pm$ 1.4 vs 4.7 $\pm$ 1.9, t=3.1, p<.005) and with their 20 age matched control subjects (6.2 $\pm$ 1.4 vs 4.8 $\pm$ 1.9, t=2.6, p<.05).

VBR values of the DSM-III schizophrenic types are reported in Fig. 2. In the comparison with the control group significantly higher values of VBR were observed in each group of patients (7.3 $\pm$ 1.6, t=3.7, p<.001 for the disorganized type; 6.6 $\pm$ 2.1, t=2.2, p< .05 for the paranoid type; 7.2 $\pm$ 2.2, t=3.2, p<.005 for the undifferentiated type and 8.0 $\pm$ 3.0, t=3.9, p<.001 for the residual type) while no significant differences were detected in the comparison between the DSM-III type groups.

Ventricles abnormally enlarged were found in 3 disorganized, in 1 paranoid, in 2 undifferentiated and in 3 residual patients.

Fig. 3 displays VBR of the 29 schizophenics grouped according to the positive-negative symptomatology. Once again each of the three groups of patients had VBR values significantly higher than the control group (7.0 $\pm$ 2.4, t=2.9, p<.01 for positive; 8.1 $\pm$ 2.6, t= 4.6, p<.001 for negative and 6.8 $\pm$ 1.6, t=3.3, p<.005 for mixed group). No differences were detected between the three groups of schizophrenics. Abnormal enlargement was observed in 3 positive, 3 negative and 3 mixed patients.

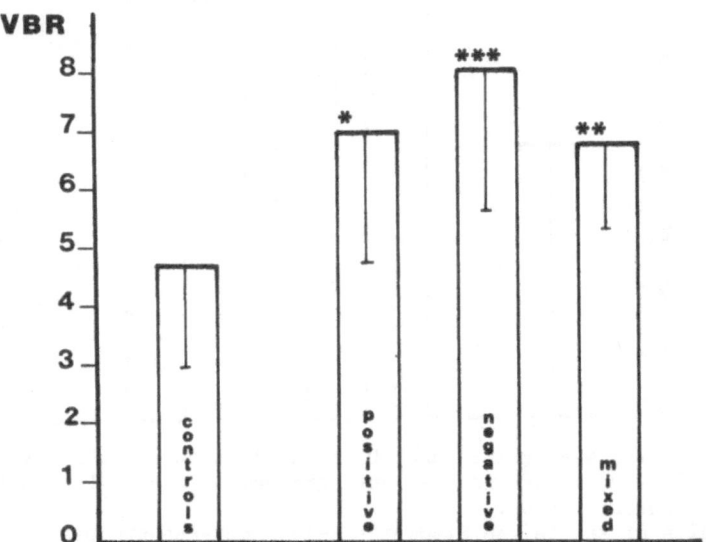

FIGURE 3   VBR mean values and SD of the 29 control subjects and of the 29 schizophrenics patient grouped according to the positive negative criteria of Andreasen (1982).
Results of two-tailed Student's t-Test between groups:
* = p<.01; ** = p<.005 and *** = p<.001 vs control group.
No significant differences between the patients groups.

As reported in Tab. 3 linear correlation analysis between VBR as independent variable and the subscales of the SANS and the SAPS performed on the whole sample of schizophrenics did not show significant results except for Anhedonia-Asociality of the SANS subscales.

Table 3.   Linear correlation analysis between SANS and SAPS
           subscales and VBR on the whole schizophrenic sample.

SANS subscales	Mean value ± S.D.	correlation coefficients	p value
Affective flattening or blunting	3.28 ± 1.22	.01	N.S.
Alogia	3.24 ± 1.46	.13	N.S.
Avolition-apathy	3.83 ± 1.20	.16	N.S.
Anhedonia-asociality	4.28 ± 1.07	.39	<.05
SAPS subscales			
Hallucinations	2.10 ± 2.09	- .19	N.S.
Delusions	3.14 ± 1.55	- .08	N.S.
Bizarre behavior	2.55 ± 1.68	.05	N.S.
Positive formal thought disorder	4.28 ± 1.07	- .21	N.S.

COMMENT

Data collected by this investigation seem to indicate that young, non institutionalized schizophrenics tend to have cerebral ventricles larger than age matched healthy subjects. In these schizophrenics VBR was normally distributed and it was not related with the duration of illness.

In accordance with most of the studies showing that lateral brain ventricles are abnormally enlarged in many schizophrenics, VBR values which have been considered by some AA [2,4] suggestive of a CSN abnormality were found in some patients but not in any control subjects. We are aware that this assessment can be weakened by the procedures adopted in this study for collecting the control subjects; in particular we cannot rule out that the selection of the wide control sample, where control subjects of this investigation were picked out, may be done also on the basis of brain CT scan data and not only on clinical data. Two main instances however support the reliability of these findings: the control group has a normal distribution of VBR with a close variability to that of schizophrenics; moreover VBR was higher in the schizophrenics than in the control subjects even when the 9 patients with abnormal VBR values were excluded from the comparison.

This investigation also indicates that the DSM-III types of schizophrenic disorder do not show different VBR, even though we cannot provide any data for the catatonic type. Similarly there are no differences between the patients with a predominance of positive vs negative or mixed symptoms. Although abnormal VBR was more likely to be found among the patients of the DSM-III residual type they were also detected in all the other groups. Moreover no significant correlation was found between VBR and schizophrenic clinical features.

These data therefore do not confirm the relationship between abnormal ventricular enlargement and negative symptoms currently reported in literature [16,17]; they rather indicate that ventricular enlargement is widely spread among the schizophrenics. These preliminary findings cannot provide evidence that ventricular enlargement is specifically related with schizophrenic disorder. The specificity of this brain structural evidence needs to be confirmed through systematic comparisons with other psychiatric disorders.

At the present stage of this study no data are available to verify the relations of this CT parameter with neuropsychological performances, treatment response and long-term outcome which will be achieved through the future development of the investigation.

REFERENCES

1.   Johnstone, EC, Crow, TJ, Frith, CD, Husband, J and Kreel L (1976). Cerebral Ventricular Size and Cognitive Impairment in Chronic Schizophrenia. Lancet, 924-926
2.   Weinberger, DR, Torrey, EF, Neophitides, AN and Wyatt RJ (1979). Lateral Cerebral Ventricular Enlargement in Chronic Schizophrenia. Arch Gen Psychiatry, 36, 735-739
3.   Golden, CJ, Moses, JA, Zelazowski, R, Graber, B, Zatz, LM, Horvath T and Berger A (1980). Cerebral Ventricular Size and Neuropsychological Impairment in Young Chronic Schizophrenics. Arch Gen Psychiatry, 37, 619-623
4.   Andreasen, NC, Smith, MR, Jacoby, CG, Dennert, JW and Olsen SA (1982). Ventricular Enlargement in Schizophrenia: Definition and Prevalence. Am J Psychiatry, 139, 292-296
5.   Nasrallah, HA, Jacoby, CG, McCalley-Whitters, M and Kuperman S (1982). Cerebral Ventricular Enlargement in Subtypes of Chronic Schizophrenia. Arch Gen Psychiatry, 39, 774-777
6.   Nybach, H, Wiesel, FA, Berggren, BM and Hindmarsch, T (1892). Computed Tomography of the Brain in Patients with Acute Psychosis and in Healthy Volunteers. Acta Psychiatr Scand, 65, 403-414
7.   Okasha, A and Madkour O (1982). Cortical and Central Atrophy

in Chronic Schizophrenia: A Controlled Study. Acta Psychiat Scand, 65, 29-34

8. Weinberger, DR, DeLisi, LE, Perman, GP, Targum, S and Wyatt RJ (1982). Computed Tomography in Schizophreniform Disorder and Other Acute Psychiatric disorders. Arch Gen Psychiatry, 39, 778-783

9. Dewan, MJ, Pandurangi, AK, Howard, LEE, S, Ramachandran, T, Levy, B, Boucher, M, Yozamitz, A and Major L (1986). A Comprehensive Study of Chronic Schizophrenic Patients. Acta Psychiat Scand, 73, 152-160

10. Nasrallah, HA, Kuperman, S, Hamra, BJ and McCalley-Whitters M (1983). Differences Between Schizophrenic Patients With and Without Large Cerebral Ventricles. J Clin Psychiatry, 44, 407-409

11. Schulz, SC, Sinicrope, P, Kishore, P and Friedel RO (1983). Treatment Response and Ventricular Brain Enlargement in Young Schizophrenic Patients. Psychopharmacol Bull, 19, 510-512

12. Revelay, MA (1985). Ventricular enlargement in schizophrenia: the validity of Computerized Tomographic Findings. Br J Psychiatry, 147, 233-240

13. Nasrallah, HA, Olson, SC, McCalley-Whitters, M, Chapman, S and Jacoby CG (1986). Cerebral Ventricular Enlargement in Schizophrenia. Arch Gen Psychiatry, 43, 157-159

14. Cazzullo, CL, Sacchetti, E, Vita, A, Illeni, MT, Bellodi, L, Maffei, C, Alliati, A, Bertrando, P, Calzeroni, A, Ciussani, S, Conte, G, Pennati, A and Invernizzi, G (1985). TAC Cerebral Ventricular Size in Schizophrenic spectrum disorders. Relathionship to clinical neuropsychological and immunogenetic variables. In: Shagas, C et Al (eds.) "Biological Psychiatry". p.1060. (Amsterdam: Elsevier Science Publisher).

15. Kemali, D, Mai, S, Galderisi, S, Ariano, MG, Cesarelli, M, Milici, N, Salvati, A, Valente, A and Valpe M (1985). Clinical and Neuropsychological correlates of Cerebral Ventricular Size. J Psychiatr Res, 19, 587-596

16. Andreasen, NC, Olsen, SA, Dennert, JW and Smith, MR (1982). Ventricular Enlargement in Schizophrenia: Relationship to Positive and Negative Symptoms. Am J Psychiatry, 139, 297-302.

17. Williams, AO, Reveley, MA, Kolakowska, T, Ardern, M and Mandelbrote BM (1985). Schizophrenia with Good and Poor Outcome: II. Cerebral Ventricular Size and its Clinical Significance. Br J Psychiatry, 146, 239-246

18. Weinberger, DR, Bigelow, LB, Kleinman, JE, Klein, ST, Rosenblatt, JE and Wyatt RJ (1980). Cerebral Ventricular Enlargement in Chronic Schizophrenia: Association with a Poor Response to Treatment. Arch Gen Psychiatry, 37,11-13

19. Donnelly, EF, Weinberger, DR, Waldman, IN and Wyatt RJ (1980). Cognitive Impairment Associated with Morphological Brain Abnormalities on Computed Tomography in Chronic Schizophrenic Patients. J Nerv Ment Disease, 168, 305-308

20. Trimble, M and Kingsley, D (1978). Cerebral Ventricular Size in Chronic Schizophrenia. Lancet, 278-279

21. Tanaka, Y, Hazama, H, Kawakara, R and Kobayashi K (1981). Computerized Tomography of the Brain in Schizophrenic Patients: A

Controlled Study. Acta Psychiat Scand, 63, 191-197
22. Jernigan, TL, Zatz, LM, Moses, JA and Berger PA (1982). Computed Tomography in Schizophrenics and Normal Volunteers: I. Fluid Volume. Arch Gen Psychiatry, 39, 765-770
23. Shima, S, Kanba, S, Masuda, Y, Tsukumo, T, Kitamura, T and Asai M (1985). Normal Ventricles in Chronic Schizophrenics. Acta Psychiatr Scand, 71, 25-29
24. American Psychiatric Association (1980). DSM-III : Diagnostic and Statistical Manual of Mental Disorders. 3rd ed, Washington, DC, American Psychiatric Association
25. Spitzer, R and Williams, JBW (1985). Structured Clinical Interview for DSM-III-Psychotic Disorders Version. Biometrics Research Department, New York State Psychiatric Institute, New York
26. Andreasen, NC (1981). The Scale for the Assessment of Negative Symptoms (SANS). Iowa city, Iowa: University of Iowa.
27. Andreasen, NC (1984). The Scale for the Assessment of Positive Symptoms (SAPS). Iowa city, Iowa, University of Iowa
28. Andreasen, NC and Olsen SA (1982). Negative vs Positive Schizophrenia: Definition and Validation. Arch Gen Psychiatry, 39, 789-794.
29. Freund, RJ and Littel, RC (eds) (1981): SAS for Linear Models. A guide to the ANOVA and GLM Procedures. SAS Institute Inc. North Carolina.

5
Hippocampal metabolic function in schizophrenia

C.A. Tamminga, K. Lyons, S.K. Kuo and G.K. Thaker

INTRODUCTION

A role for the mammalian hippocampus and other limbic structures in generating symptoms of schizophrenia has been frequently suggested. The hippocampus is richly connected to important areas of neocortex, including the prefrontal granular cortex, the temporal neocortex, and the parietal areas. It is said to be functionally involved in a number of nonmotor behaviors, chiefly attention and motivation, and in the integration between informational inputs and motivational orientation. The traditional association of the hippocampal structures to short-term memory and to memory-storage pathways provides the hippocampus with functional pathways for memory retrieval which, if pathologically functioning, might be a mechanism for sensory hallucinations. Psychotic symptoms do occur in humans with phencyclidine injection, a drug known to potently activate the hippocampus. And, descriptions of various sorts of hippocampal pathology have been made in schizophrenia.

The hippocampus is a gyrus of the allocortex located on the ventricular aspect of the temporal lobe. Its histologic structure is more primitive than neocortex, and as such, it at one time was considered simpler to understand. Alas, a recent quote from Dr. Weiskrantz suggests the opposite: "The striking aspect of the hippocampus is the anatomic elegance of its structure, but in contrast there is really appalling ignorance about what this elegance means" [1].

The primary input pathway to the hippocampus derives from the entorhinnal cortex. It is through this entorhinnal pathway that the hippocampus receives most of its rich and diverse innervation, because this latter structure receives afferents from the temporal neocortex, the prefrontal granular cortex and the prepiriform and periamygdala cortex, and, also, from rostrally directed brain stem pathways, from the locus caeruleus and dorsal raphe nucleus. In addition to the entorhinnal area, the anterior thalamic nucleus, the septum, and hypothalamus contribute afferents to hippocampus [1].

The hippocampal output pathways are multiple and each may well act independent of the other to subserve different CNS-mediated functions. The efferents directly innervate the anterior ventral thalamus, cingulate cortex, the septum and hypothalamus; also, the efferents extend caudally to the pontine nuclei, and periaquaductal grey. From what is known about the anatomic connections of the hippocampus, it is clear that this structure is complexly connected with many other brain areas, directly with limbic structures and septum, but indirectly with all areas of brain; and, it is intricately interconnected within its own structures [1].

Already, anatomic abnormalities in the hippocampus of schizophrenia patients have been reported. The area of the hippocampal cortex and the amygdala are reduced [2]. In addition, a reduction in the size of the parahippocampal cortex has been observed with postmortem material and an increase in proximal lateral ventricular size [3]. Schibel at one time found hippocampus pyramidal cell rearrangement in twelve subjects with schizophrenia [4]. More recently, however, in an attempted replication of that work, no significant cellular rearrangement was found, but an association between the degree of pyramidal cell disorganization and the severity of the psychosis [5].

We have evaluated neuronal activity in the hippocampus and its related structures in schizophrenia using two different methods. In one approach we made metabolic measurements in hippocampal areas in an animal model of the illness using phencyclidine in rodents and quantitative glucose autoradiography. In the other method we explore the same question by carrying out metabolic measurements in the hippocampus and related brain areas in nonmedicated schizophrenic patients with active, psychotic symptoms. The assessment of neuronal activity in both paradigms was carried out using tracer-labled 2-deoxyglucose (2DG) accumulation (^{14}C 2DG for rats and ^{18}F 2DG for humans) with autoradiography for the animal tissue [6] and positron emission tomography PET for the human subjects [7]. The schizophrenic patients were neuroleptic-free for longer than 4 weeks and displayed active but stable psychotic symptoms.

HIPPOCAMPAL SYSTEM METABOLIC RESPONSE TO PCP IN THE RAT

Phencyclidine is a psychotomimetic drug which produces psychotic symptoms in humans characteristic of an acute schizophrenic reaction with delusions, thought disorder, and paranoia. It causes symptoms more similar to schizophrenia, than any of the other psychotomimetic compounds. It stimulates a distinct receptor in brain, now called the PCP-receptor, for which there is thought to exist an endogenous peptide ligand ("endopsychosin"). This PCP receptor has a distinct distribution with high densities in the hippocampus and cortex, and is thought to be inversely functionally linked with one of the glutamate receptors, the N-methyl-d-aspartate (NMDA) receptor. Thus, there is reason to believe that phencyclidine stimulates parts of the brain which are associated with psychosis, and that the hippocampus may well be important in this psychotogenic action.

Quantitative glucose metabolism in the rat brain was assessed in the usual manner [8]. A complete description of the method has been previously reported [6]. In short, ^{14}C-2-deoxyglucose (1.25 mCui/kg) was administered to each of 10 animals, who were pretreated with phencyclidine or saline. After a 45 min. incorporation period, the animals were sacrificed, their brains sectioned and autoradiographed, and optical density of the ^{14}C-2DG read on an IBAS II Image Analysis System. Data was analyzed using analysis of variance with post hoc tests for significance.

A single dose of PCP (4 mg/kg) increased glucose utilization in the pyramidal cell layer of the rat hippocampus. Other cell layers in the hippocampus proper were not altered. But, some of the structures in the associated hippocampal areas evidenced striking metabolic stimulation; these areas included the subiculum and entorhinnal cortex. Also, some of the primary projection areas of the hippocampal complex evidenced dramatic PCP changes, namely the anteroventral thalamus and the anterior cingulate cortex (Table 1). Other hippocampal projection areas were not affected by PCP, including the large projection to the lateral septum and the hypothalamus (Table I).

These data suggest that PCP selectively alters glucose metabolism within the hippocampus and its thalamocortical efferent pathway, but not in its septal-hypothalamic projection. Perhaps the neurons in the hippocampus responsible for higher cortical projections, as opposed to hypothalamic-mediated functions, selectively bear PCP receptors, or are sensitive to exogenous PCP. Speculatively, activation of the endogenous PCP-receptor ligand, now called endopsychosin [9], could selectively activate this higher-cortical hippocampal pathway, and could thereby "endogenously" generate psychotic symptoms in schizophrenic illness.

HIPPOCAMPAL SYSTEM METABOLIC RESPONSE TO SCHIZOPHRENIA IN MAN

The hippocampus, and its afferent brain areas (parahippocampal cortex, dorsomedial prefrontal cortex, and periamygdala cortex) and its projection fields (anterior cingulate cortex) were examined using FDG/PET in 12 drug-free, DSM III-diagnosed, schizophrenic patients, compared with 12 age and sex matched controls. PET scans were carried out on a NeuroPET scanner at the NINCDS, which has a stated resolution of 6 mm, to allow visualization of lower cortical and subcortical structures, as well as the cortical mantel. Areas selected for measurement in the hippocampal system were compared between normal controls and schizophrenic patients using t statistic, with respect to regional metabolic activity at rest. Some pertinent brain areas functionally linked with the hippocampus could not be clearly enough distinguished on these human PET scan for metabolic analysis, like the locus caeruleus, raphe nucleus, anteroventral thalamus, subiculum or septal areas.

Glucose utilization in the human hippocampus and its related structures was measured as absolute metabolic rate and relative rate. The former were calculated directly from the positron scanner data and the tracer plasma curves according to the operational equation of the method modified for human application [10]. Relative values were calculated from individual area metabolic values divided by that individual's cerebellar metabolic value. Metabolic values for glucose in the hippocampus, both the absolute and the relative values, showed a decrease in schizophrenic subjects compared with normals (Table 2). Whereas the prefrontal projection demonstrated no differences in the two groups, the amygdala did probably demonstrate some modest hypometabolism in the schizophrenic group. A chief efferent projection area of the hippocampus, the anterior cingulate cortex, also showed a significant diminution in metabolism in schizophrenics (Table 2). The area of thalamus assessed so far in this human group is the mid-central thalamic area (not anterior or ventral) and it shows no change. It is important to note that cerebellar values show no difference between the two groups.

Each of the schizophrenic patients was tested with PET/FDG after at least 4 weeks neuroleptic free, and when they were stably psychotic. It has been our observation that most schizophrenics, upon neuroleptic withdrawal, pass through a behaviorally and psychologically volatile period for 2-3 weeks, then settle in to a highly psychotic yet nonvolatile and nonactivated state. It is in such a "stable" state that these schizophrenic patients were tested, both with the Brief Psychiatric Rating Scale (BPRS) for symptom assessment and the PET/FDG for metabolic assessment. The schizophrenic subject and their controls were young and, the patients, highly psychotic. We looked at correlations within this patient group, albeit a very small population, between metabolism and psychosis (BPRS, factor I). We found no relationship between level of psychotic symptoms and magnitude of glucose metabolism in these areas.

CONCLUSION

We suggest the existance of limbic system alterations with schizophrenia, based on the study of the PCP-treated rat, and on the glucose metabolism changes in some of those same brain areas in actively psychotic patients. It must be emphasized that these clinical data are only preliminary, in that there is a small subject number and, as yet, an incomplete analysis. But the results of the human study parallel the changes with PCP in rat with respect to localization, making these clinical data interesting. Although the animal model and the human PET data overlap with respect to brain areas altered, the direction of metabolic change is opposite in directions. PCP elevates glucose uptake in limbic structures while schizophrenia reduces the uptake. Assuming that these two entitites are related and both related to the metabolic changes, differences between rat and human brain or between PCP treatment and schizophrenia may play a role in the directional difference.

Based on these data and on the obviously important role of the hippocampus in human behavior, this region deserves further study in schizophrenia. Perhaps only selected disease subtypes or certain symptom clusters may be linked with hippocampal dysfunction. Or perhaps, the hippocampus may be a part of a critical psychosis circuit where it has a direct or indirect role. Hence, the pathologic alterations may vary. The confluence of multiple sophisticated methodologies and areas of new knowledge make it possible to begin to examine the role of particular brain regions in schizophrenia. Additional knowledge will make these approaches even more viable. Such attention to the hippocampus in the future will be undoubtedly interesting.

Table 1.   PCP Action on Hippocampal Structures in the Rat*

Hippocampal Afferent Areas:	PCP (N=5)	Saline (N=5)
Entorhinnal Cortex*	116.0 ± 7.2	88.0 ± 8.6
Dorsolateral Prefrontal Cortex	136.4 ± 14.8	125.6 ± 5.6
Dorsal Raphe	111.5 ± 10.4	106.8 ± 4.4
Locus Ceruleus	110.1 ± 9.6	84.6 ± 3.1
Hippocampal Efferent Areas:		
Subiculum*	213.4 ± 46.9	104.0 ± 6.0
Lateral Septum	88.7 ± 12.7	81.3 ± 6.4
Anteroventral Thalamus*	208.7 ± 32.3	121.1 ± 7.3
Anterior Cingulate Cortex*	216.6 ± 34.7	131.5 ± 7.7
Nucleus Accumbens*	199.1 ± 24.2	131.1 ± 9.1
Periaquaductal Grey	107.0 ± 11.9	96.1 ± 4.6
Hypothalamus	84.1 ± 7.2	94.9 ± 13.8

* umol glucose/100 gm tissue/min; mean (SD)

*$p < .05$

Table 2. Hippocampus and Related Structures*

	Absolute Values			Relative Values		
	Schizo	NL	P	Schizo	NL	P
Dorsomedial Prefrontal Cortex	10.6 (2.1)	11.8 (2.6)	NS	1.4 (.30)	1.5 (.30)	NS
Cingulate Cortex	9.4 (1.9)	12.5 (3.3)	<.01	1.2 (.26)	1.6 (.36)	<.02
Hippocampus and Parahipp. Cortex	6.9 (1.6)	9.2 (2.1)	<.007	0.91 (.20)	1.1 (.29)	<.05
Amygdala Area	6.0 (1.1)	7.0 (1.4)	=.08	0.78 (.10)	0.87 (.09)	<.05
Thalamus (mid central)	10.0 (1.9)	11.1 (3.1)	NS	1.3 (.35)	1.3 (.30)	NS
Cerebellum	7.8	8.0	NS	-	-	-

* mg glucose/100 gm tissue/min; N=12/group; left hemisphere values; mean SD

** area metabolism/cerebellum metabolism

References

1. Isacson, RL, and Pribram, KH (eds.) (1975). Hippocampus, (Plenum Press, New York).

2. Bogerts, B, Meertz, E and Schonfeldt-Bausch, R (1985). Basal ganglia and limbic system pathology in schizophrenia. Arch Gen Psychiatry, 42:784-791.

3. Brown, R, Colter, N, Corsellis, J, Crow, TJ, Frith, CD, Jagore, R, Johnstone, EC and Marsh, L (1986). Postmortem evidence of structural brain changes in schizophrenia, differences in brain weight, temporal horn area and parahippocampal gyrus compared with affective disorder. Arch Gen Psychiatry, 43:36-42.

4. Scheibel, AB and Kovelman, JA (1981). Disorientation of the hippocampal pyramidal cell and its processes in the schizophrenic patient. Biol Psychiatry, 16:101-102.

5. Altshuler, LL, Conrad, A, Kovelman, JA and Scheibal, A (1987). Hippocampal pyramidal cell orientation in schizophrenia. Arch Gen Psychiatry, 44:1094-1098.

6. Tamminga, CA, Tanimoto, K, Kuo, S, Chase, TN, Contreras, PC, Rice, KC, Jackson, AE and O'Donohue, TL (1987). PCP-induced alterations in cerebral glucose utilization in rat brain: blockade by metaphit, a PCP-receptor-acylating agent. Synapse, 1:497-504.

7. Tamminga, CA, Thaker, GK, Alphs, LD and Chase, TN (1988). Limbic system: localization of PCP drug action in rat and schizophrenia manifestations in humans. In: Schulz, SC and Tamminga, CA (eds.) "Schizophrenia: A Scientific Perspective". (Oxford Press, New York).

8. Sokoloff, L, Reivich, M, Kennedy, C, DesRosiers, MH, Patlak, CS, Pettigrew, KD, Sakurada, O and Shinohara, M (1977). The [^{14}C] deoxyglucose method for the measurement of local cerebral glucose utilization: theory, procedure, and normal values in the conscious and anesthetized albino rat. J Neurochem, 28:897-916.

9. Monahan, JB, Contreras, PC, Lanthorn, TH, DiMaggio, DA, Handelmann, G, Pullan, LM, Gray, NM and O'Donohue, TL (1988). The phencyclidine receptor complex: interaction with excitatory amino acids and endogenous ligands. In: Schulz, SC and Tamminga, CA (eds.) "Schizophrenia: A Scientific Perspective". (Oxford Press, New York).

10. DiChiro, G, Brook, RA, Patronas, NJ, Bairamian, D, Kornblith, PL, Smith, BH, Mansi, L and Barker, J (1984). Issues in the in vivo measurement of glucose metabolism of human central nervous system tumors. Ann Neurol, 15:138-146.

SECTION 2:
AFFECTIVE DISORDERS: THE IMPACT OF NEUROIMAGING TECHNIQUES

6
CT and MRI findings in affective disorders: clinical and research implications

H.A. Nasrallah, J.A. Coffman and S.C. Olson

INTRODUCTION

Over the past decade, new brain imaging technologies such as computerized tomography (CT) and magnetic resonance (MR) have provided psychiatric researchers an unprecedented tool to study the in vivo structural integrity of the brain in psychiatric disorders. Most of the initial reports focused on schizophrenia, and the various structural abnormalities detected were at first believed to be possible biological markers for schizophrenia. However, similar findings were soon reported in patients with affective disorders and other psychiatric conditions as well, indicating that neuroanatomical abnormalities such as cortical atrophy or ventricular enlargement are not specific to any particular disorder.

However, further research into the clinical and biological correlates of structural brain abnormalities in both schizophrenia and affective disorders indicates that certain CT or MR brain imaging findings may be useful in identifying subtypes of the illness characterized by a cluster of symptoms or the presence of concomitant biochemical abnormalities. Thus, despite the general nonspecificity of the structural abnormalities, some applications for clinical management and research strategies have emerged.

In this paper, an overview of the CT literature and the more recent MR findings in affective disorders (bipolar and unipolar) is presented, and the various reports of clinical and biological correlates of these findings are reviewed. Current applications of these findings and future research trends in this area are then presented and· discussed.

CT STUDIES

1. <u>Cerebral Ventricular Enlargement</u>: (Table 1) The lateral
   cerebral ventricles to brain ratio (VBR) in manic patients
   was found to be enlarged (Pearlson and Veroff 1981,
   Johnstone, et. al. 1981, Nasrallah, et. al. 1982, Rieder et.
   al. 1983, Weinberger, et. al. 1983, and Luchins, et. al.
   1984). However, Tanaka, et. al. (1982) found enlargement
   only in the third ventricles in manics compared to controls.
   Owens, et. al. (1985) found VBR in manics to be between
   schizophrenics and neurotic controls and no different from
   either. In depressed patients, Jacoby and Levy (1981) found
   ventricular enlargement in elderly depressives compared to
   matched controls, Targum, et. al. (1983) and Scott, et. al.
   (1983) reported larger ventricles in delusional depression
   and Dolan, et. al. (1985) found large ventricles in
   depressed patients. Schlegel and Kretzschmar (1987) found no
   difference in VBR between affectively ill patients and
   controls, but linear measurements were enlarged in the
   patient group. Older, male, psychotic unipolar, DST non-
   suppressors and lithium treated patients accounted for the
   differences with controls.

Table 1. **VBR VALUES IN CT STUDIES OF AFFECTIVE DISORDERS**

	VBR (%)	
	<u>AD</u>	<u>C</u>
Pearlson and Veroff (1981)	6.5	3.6
Johnstone, et. al. (1981)	11.5	10.2 NS
Nasrallah, et. al. (1982)	7.5	3.2
Weinberger, et. al. (1982)	3.8	3.0 NS
Rieder, et. al. (1983)	5.3	3.8 NS
Scott, et. al. (1983)	9.5	4.2
Targum, et. al. (1983)	5.1	2.9
Cazzullo, et. al. (1984)	6.3	
Pearlson, et. al. (1984)	6.6	4.5
Luchins, et. al. (1984)	4.5	3.0
Shima, et. al. (1984)	11.2	9.1
Dolan, et. al. (1985)	7.2	5.6
Pearlson, et. al. (1985)	8.8	4.5
Kolbeinsson, et. al. (1986)	10.3	8.4
Schlegel and Kretzschmar (1987)	7.4	6.9 NS

Table 2 summarizes three studies of third ventricular size in affective patients.

**Table 2.   CT STUDIES OF 3RD CEREBRAL VENTRICULAR
SIZE IN AFFECTIVE DISORDERS**

	Length (mm)		
	AD	C	
Tanaka, et. al. (1982)	7.9	7.6	NS
Nasrallah, et. al. (1985)	4.1	3.4	NS
Schlegel·and Kretzschmar (1987)	4.3	3.4	p <.004

2. Sulcal Widening: (Table 3) Cortical surface atrophy have also been reported in some manic patients (Nasrallah, et. al. 1981, Pearlson and Veroff 1981, Nasrallah, et. al. 1982, and Rieder, et. al. 1983). Schlegel and Kretzschmar (1987) found no difference in the mean width of the interhemispheric fissure and the sylvian fissure between affective patients and controls.

**Table 3.   CT STUDIES OF CORTICAL (SULCAL) ATROPHY
IN AFFECTIVE DISORDERS**

	% Abnormal	
	AD	C
Pearlson and Veroff (1981)	12.5%	
Nasrallah, et. al. (1982)	25.0%	3.7%
Tanaka, et. al. (1982)	58.0%	54.0% frontal
	75.0%	50.0% temporal
	40.0%	70.0% parietal
	25.0%	0.4% occipital
Rieder, et. al. (1983)	22.0%	23.0% (schizo)
Dolan, et. al. (1986)	greater in frontal & temporal	
Kolbeinsson, et. al. (1986)	greater in AD vs C	

3. Cerebellar Atrophy: (Table 4) Evidence of cerebellar atrophy in some manic patients (without a history of alcohol abuse) was also reported by several investigators (Nasrallah, et. al. 1981, Pearlson and Veroff 1981, Nasrallah, et. al. 1982, Lippmann, et. al. 1982, Weinberger, et. al. 1982, and Rieder, et. al. 1983). Nasrallah, et. al. (1982) found larger VBR, but not more sulcal widening in manic patients with cerebellar atrophy.

**Table 4.   CT STUDIES OF CEREBELLAR ATROPHY
IN AFFECTIVE DISORDERS**

	% Abnormal	
	AD	C
Nasrallah, et. al. (1981)	27%	3%
Pearlson, et. al. (1981)	0%	0%
Nasrallah, et. al. (1982)	21%	4%
Lippmann, et. al. (1982)	17%	5%
Weinberger, et. al. (1982)	9%	0%
Rieder, et. al. (1983)	11%	7% (schizo)
Yates, et. al. (1987)	8% (bi)	3%
	0% (uni)	

4.   Abnormal Cerebral Asymmetries: (Table 5)   Only three studies
of cerebral hemisphere asymmetries on CT scans in manic
patients have been reported. Weinberger, et. al. (1982)
found no significant reversals in affective patients compared
to psychiatric and medical controls. Tanaka, et. al. (1982)
reported increased reversal of normal cerebral asymmetries in
bipolar affective patients vs controls. Tsai and Nasrallah
(1983) found significantly less reversals in manic males
compared to schizophrenic males.

**Table 5.   CT STUDIES OF ABNORMAL CEREBRAL ASYMMETRIES
IN AFFECTIVE DISORDERS**

	Frequency of Reversals
Weinberger, et. al. (1982)	No difference vs controls
Tanaka, et. al. (1982)	More reversal vs controls
Tsai and Nasrallah (1983)	Less reversal vs schizo
Dewan, et. al. (1987)	No difference vs controls

5.   Density Deficits: Coffman and Nasrallah (1984) reported
significantly higher frequency of decreased left hemisphere
signal intensity values in manics and schizophrenics compared
to controls.   Jacoby, et. al. (1981) found that brain tissue
density in elderly depressives was more similar to dementia
than matched healthy controls, and that decreased density in
depressives was associated with large VBR.   Schlegel and
Kretzschmar (1987) found lower attenuation values in
psychotic vs non-psychotic affective patients, lower values
in men than in women, lower values in lithium treated than
non-lithium treated, and lower values in unipolar compared to
bipolar patients.   They found no correlation between age and
density in patients or controls.

Several clinical and biological correlates of ventricular enlargement and cortical atrophy have been reported. The following is a summary of those findings:

Clinical Correlates of Large VBR in Affective Disorder

a) later age of onset (Jacoby and Levy 1980)

b) presence of delusions and hallucinations (Pearlson and Veroff 1981, Targum, et. al. 1983, Luchins, et. al. 1984)

c) lower IQ scores (Targum, et. al. 1983)

d) frequent hospitalizations and frequent unemployment (Pearlson, et. al. 1984)

e) higher mortality in elderly depressives (Jacoby, et. al. 1981)

f) fewer episodes of illness per year (Nasrallah, et. al. 1984)

g) severity of illness (Standish-Barry, et. al. 1982)

Biological Correlates of Large VBR in Affective Illness

a) higher urinary cortisol (Kellner, et. al. 1983)

b) no association with dexamethason suppression (Schlegel and Kretzschmar 1987)

c) association with hypothyroidism but not with lithium intake (Johnstone, et. al. 1986)

d) decreased plasma free tryptophan (Standish-Barry, et. al. 1986)

e) increased CSF 5-HIAA (Standish-Barry, et. al. 1986)

f) no relationship to CSF HVA (Standish-Barry, et. al. 1986)

Clinical Correlates of Cortical Atrophy in Affective Disorder

a) association of frontal atrophy with ECT (Calloway, et. al. 1981)

b) frontal and temporal atrophy with ECT (Dolan, et. al. 1986)

c) no relationship to ECT (Nasrallah, et. al. 1984; Kolbeinsson, et. al. 1986)

d) related to duration of illness (Kolbeinsson, et. al. 1980)

Clinical ᴄorrelates of Brain Tissue Density in Affective Disorder

a)   lower values in: psychotic vs non-psychotic
                        men vs women
                        lithium treated vs non-lithium treated
                        unipolar vs bipolar

b)   higher values associated with small VBR

c)   no association with dexamethasone response or age (Schlegel
     and Kretzschmar, 1987)

d)   no relationship to VBR or sulcal atrophy (Coffman and
     Nasrallah, 1985)

MRI STUDIES

Several studies of brain MRI findings have appeared in the
literature over the past few years, mostly in schizophrenic
patients. A few studies were done in affective patients and are
summarized below:

1.   Cerebral Ventricular Enlargement: Besson, et. al. (1987)
     reported no differences on MRI in VBR between bipolar
     patients vs controls. Nasrallah, et. al. (1987) also found
     no ventricular enlargement on MRI scans in bipolar patients,
     and no differences between bipolars, schizophrenics and
     controls.

2.   Cerebral Cranial and Frontal Lobe Atrophy: Nasrallah, et.
     al. (1987) found no decrease in cerebral and cranial size in
     bipolar patients as was reported by Andreasen, et. al. (1986)
     in schizophrenic patients. No differences were also found in
     frontal lobe size between bipolar patients and controls.

3.   Cerebellar Atrophy: Nasrallah, et. al. (1987) found no
     decrease in cerebellar size between bipolar patients and
     controls.

4.   Corpus Callosum Size: No differences were found in corpus
     callosum length, area and thickness measures between bipolar
     patients and controls (Nasrallah, et. al. 1987).

5.   T1 and T2 Relaxation Times: Rangel-Guerra, et. al. (1982)
     reported that T1 values in the temporal and frontal regions
     were higher in bipolar patients than in controls, and that
     after lithium treatment, T1 values decreased to the normal
     range. Rosenthal, et. al. (1986) reported similar T1
     findings in the red blood cells of bipolar patients. On the
     other hand, Knowles (1985) reported that lithium therapy
     increases T1 in bipolars to above the level of controls,
     while Besson, et. al. (1987) found that T1 values of both
     bipolar patients and normal controls are the same.

As for clinical correlates of MRI findings in bipolar patients, there are very few such reports in the literature. Nasrallah, et. al. (1987) found a strong correlation between the presence of first degree family history of psychiatric hospitalization and smaller cerebral size in both bipolars and schizophrenics. They also found no relationship between perinatal complications and MRI measures.

The only biological correlate of MRI brain finding in bipolar patients is a study by Schwarzkopf, et. al. (1987) showing a significant negative correlation between visual evoked potential amplitudes and frontal and cerebral size.

DISCUSSION

It is clear that much more investigation remains to be done in the area of structural brain imaging in affective disorders. For example, there are very few controlled CT or MRI studies in unipolar major depression, especially the younger (under 50 years) age groups. The few reports regarding clinical or biological correlates of neuroanatomical abnormalities in affective illness are isolated findings and not methodologically consistent.

However, despite the general nonspecificity of the CT and MRI findings in affective disorders, the various clinical and biological parameters which have been reported with the structural brain parameters seem to suggest some useful applications. Clinical characteristics of patients with large VBR or cortical atrophy may not only assist in improving clinical management, but may help shed light on the pathophysiology of affective symptomatology. As for research implications, the clinical and biological parameters may also assist in defining subtypes of unipolar and bipolar affective illness, and to generate new hypotheses and research strategies.

Finally, CT and MRI are very important diagnostic tools for detecting cases of "symptomatic" affective disorders secondary to neurological lesions such as tumors, infarcts, infections, trauma, degeneration, demyelination, etc., and thus avoid delaying more appropriate clinical management.

REFERENCES

1. Andreasen NC, Nasrallah HA, Dunn V, Ehrhardt JC, Grove WM, Olson SC, Coffman JA, Crossett TH. Structural abnormalities in the frontal system in schizophrenia: A magnetic resonance imaging study. Archives of General Psychiatry 43:136-144, 1986.

2. Besson JAO, Henderson JA, Foreman I, Smith FW. An NMR study of lithium responding manic depressive patients. Mag Res Imag 5:273-277, 1987.

3. Coffman JA, Nasrallah HA. Brain density patterns in schizophrenia and mania. Journal of Affective Disorders 6:307-315, 1984.

4. Jacoby RJ, Levy R. Computerized tomography in the elderly, 2. Senile dementia, diagnosis and functional impairment. Br J Psychiatry 136:256, 1980.

5. Jacoby RJ, Levy R, Bird JM. Computer tomography and the outcome of affective disorder: A follow-up of elderly patients. Br J Psychiatry 139:288, 1981.

6. Johnstone EC, Owens DGC, Crow TJ, et al. A CT study of 188 patients with schizophrenia, affective psychosis and neurotic illness. In C. Perris, G. Struwe, B. Jansson (Eds), Biological Psychiatry. Amsterdam: Elsevier/North Holland Biomedical Press, pp. 237-240, 1981.

7. Johnstone EC, Owens DGC, Crow TJ, Colter N, Lawton A, Jagoe R, Kreel L. Hypothyroidism as a correlate of lateral ventricular enlargement in manic-depressive and neurotic illness. British Journal of Psychiatry 148:317-321, 1986.

8. Kellner CH, Rubinow DR, Gold PW et al. Relationship of cortisol hypersecretion to brain CT scan alters in depressed patients. Psychiatry Research 8:191-197, 1983.

9. Knowles R. The effect of lithium on cerebral T_1 valves. Soc of Mag Res in Med Third annual meeting. August 13-17, New York p. 425, 1985.

10. Lippmann S, Manshadi M, Baldwin H, Drasin G, Rice J, Alrajeh S. Cerebellar vermis dimensions on computerized tomographic scans of schizophrenia and bipolar patients. Am J Psychiatry 139:667, 1982.

11. Luchins DJ, Lewine RRJ, Melzer HY. Lateral ventricular size, psychopathology, and mediention response in the psychoses. Biol Psychiatry 19:20, 1984.

12. Nasrallah HA, Jacoby CG, McCalley-Whitters M. Cerebellar atrophy in schizophrenia and mania. Lancet 1:1102, 1981.

13. Nasrallah HA, McCalley-Whitters M, Jacoby GG. Cortical atrophy in schizophrenia and mania: A compartive CT study. J Clin Psychiatry 43:439-441, 1982.

14. Nasrallah HA, McCalley-Whitters M, Joacoby CG. Cerebral ventricular enlargement in young manic males: A controlled CT study. J Affec Dis. 4:15-19, 1982.

15. Nasrallah HA, Jacoby CG, McCalley-Whitters M, Kuperman S. Cerebral ventricular enlargement in subtypes of chronic schizophrenia. Arch Gen Psychiatry 39:774-777, 1982.

16. Nasrallah HA, McCalley-Whitters M, Pfohl B. Clinical significance of large cerebral ventricles in manic males. Psychiatry Research 13:151-156, 1984.

17. Nasrallah HA, Olson SC, Coffman JA, McLaughlin JA, Kovach M. Cranial, cerebral and frontal size in affective disorders and schizophrenia: An MRI study. Presented at the American College of Neuropsychopharmacology Annual Meeting. December 7-11, 1987 San Juan, Puerto Rico.

18. Owens DGC, Johnstone EC, Crow TJ, Frith CD, Jagol JR and Kreel L. Lateral ventricular size in schizophrenia: Relationship to the disease process and its clinical manifestations. Psychol Med 14:27-41, 1985.

19. Pearlson GD, Verhoff AF. Computerized tomographic scan changes in manic depressive illness. Lancet 2:170, 1981.

20. Pearlson GD, Garbacz DJ, Breadey WR, Ahn HS, DePaulo JR. Lateral ventricular enlargement associated with persistent unemployment and negative symptoms in both schizophrenia and bipolar disorder. Psychiatry Res 12:1, 1984.

21. Rangel-Guerra RA, Perez-Payou I, Todd LE. Nuclear magnetic reasonance in bipolar affective disorder. Mag Res Imag 1:229, 1982.

22. Rieder RO, Mann LS, Weinberger DR, van Kammen DP, Post RM. Computed tomographic scans in patients with schizophrenia, schizoaffective, and bipolar affective disorder. Arch Gen Psychiatry 40:735, 1983.

23. Rosenthal J, Strauss A, Minkoff L, Winston A. Identifying lithium responsive bipolar depressed patients using nuclear magnetic resonance. Am J Psychiatry 143:779, 1986.

24. Schlegel S, Kretzschmar K. Computed tomography in affected
     disorders. Part 1. Ventricular and sulcal measurements.
     Biol Psychiatry 22:4-14, 1987.

25. Schlegel S, Kretzschmar K. Computed tomography in
     affective disorders. Part 2. Brain density. Biol
     Psychiatry 22:15-23, 1987.

26. Scott ML, Golden CJ, Ruedrich SL, et al. Ventricular
     enlargement in major depression. Psychiatry Res 8:91-93,
     1983.

27. Standish-Barry HMAS, Bouras N, Bridges PK, Bartlett JR.
     Pneumoencephalographic and computerized axial tomography
     scan changes in affective disorder. Br J Psychiatry
     141:614-617, 1982.

28. Standish-Barry HMAS, Bouras N, Hale AS, Bridges PK,
     Bartlett JR. Ventricular size and CSF transmitter
     metabolite concentrations in severe endrogenous depression.
     Br J Psychiatry 148:386-392, 1984.

29. Swartzkopf SB, Torello MW, Coffman JA, Olson SC, McLaughlin
     JA, Nasrallah HA. Neuroanatomical and neurophysiological
     studies in affective disorders. (In preparation).

30. Tanaka Y, Hazama H, Fukuhara T, et al. Computerized
     tomography of the brain in manic-depressive patients: A
     controlled study. Folia Psychiatr Neurol 36:137-144, 1982.

40. Targum SD, Rosen LN, Citren CM. Delusional symptoms
     associated with enlarged ventricles in depressed patients.
     South Med J 76:985-987, 1983.

50. Tsai LY, Nasrallah HA, Jacoby CG. Hemispheric asymmetries
     on computed tomographic scans in schizophrenia and mania.
     Arch Gen Psychiatry 40:1286, 1983.

51. Weinberger DR, DeLisi LE, Perman GP, et al. CT scans in
     schizophreniform disorder and other acute psychiatric
     patients. Arch Gen Psychiatry 39:778-783, 1982.

52. Weinberger DR, Luchins DJ, Morihisa JM, Wyatt RJ.
     Asymmetrical volumes of the right and left frontal and
     occipital regions of the human brain. Ann Neurol 11:97,
     1982.

53. Weinberger DR, Wagner RL, Wyatt RJ. Neuropathological
     studies in schizophrenia: A selective review.
     Schizophrenia Bulletin 9:193-212, 1983.

7
Neuromorphological correlates of mood disorders: focus on cerebral ventricular enlargement

E. Sacchetti, A. Vita, A. Calzeroni, G. Conte, F. Pollastro, A. Terzi, G. Valvassori, G. Invernizzi and C.L. Cazzullo

INTRODUCTION

The hypothesis that affective disorders or some of their subforms also might have an organic basis certainly is not new to present-day psychiatry. It may suffice to mention that Kraepelin, one of the "founding fathers of modern psychiatry" (1), suspected a degenerative process in the etiology of involutional melancholia (2).

In spite of such an influential voice, however, explicit attempts to demonstrate structural brain abnormalities in patients with affective disorders were substantially neglected for several decades. In particular, in vivo studies have been intensively focused on the pneumoencephalographic correlates of schizophrenia (3, 4), rather than on affective disorders, even though the infrequent application of this approach to them led to promising findings. For example, Nagy (5) reported clear signs of cerebral atrophy in almost a third of the manic-depressive patients he studied.

Consequently, we can readily transfer Roberts and Crow's conclusions (4) to affective disorders that "by the early '70s there was no consensus on the existence of organic damage in the brains of patients with schizophrenia", the disorder most often and best studied in neuropathologic and pneumoencephalographic terms.

In the early '70s, however, there was a specific inversion of the tendency. The advent of computerized tomography (CT) with Hounsfield (6) and of other new brain imaging technologies in rapid succession also gave psychiatry the tools to "study brain anatomy in patients in exquisite detail" (1) on a non-invasive basis and permitted a large-scale and safe screening of putative "anatomic... substrates of mental illnesses." (1).

The opening of this "new and very promising area for the in vivo morphological approach to psychiatric disorders" (7) has produced a substantial body of data that firmly and definitively supports the frequent presence of discrete neuromorphological abnormalities in schizophrenic and affective disorders (7, 8, 9, 10).

Confining ourselves to the analyses of the CT correlates of affective disorders, enlargement of both the lateral cerebral ventricles and of the third one, sulcal widening, cerebellar atrophy, abnormal cerebral asymmetries and lower brain tissue densities have been more or less consistently reported (9, 10) since Jacoby and Levy (11)first documented ventricular enlargement in a group of depressed geriatric patients.

Substantial uncertainty still persists in regard to the relevance and significance of lateral cerebral ventricular enlargement, despite the fact that this doubtlessly is the abnormality that has been most and best studied. For example, there is much data indicating that ventricular enlargement may be valuable for identifying subgroups of affective patients with well-defined clinical, pharmacological and biological characteristics (9, 10). The data, however, comes from studies that have not been sufficiently replicated, that sometimes conflict in their conclusions, and for the most part have been focused on a few selected clinical and biological variables without sufficient control of several possible sources of variation. Further, in many instances the samples that were recruited were far from representative either in numerical terms, or selection terms. Therefore, an attempt to organize the existing data into a unitary context appears to be premature. This is even more so if we consider that a number of questions concerning the heuristic value and practical relevance of ventricular enlargement still have not been answered definitively. The unresolved issues predominantly pertain to the etiopathogenetic determinants of ventricular enlargement, whether these have a causative role or only act as simple markers for a definite subtype of affective disorder. Other crucial areas of interest involve the time of appearance of ventricular enlargement and the question of whether it is static or progressive. Another problem to explore is whether this neuromorphologic abnormality, which cuts across several psychiatric disorders, is indicative of a dimensional continuum among them or represents a last common characteristic triggered by mechanisms that are at least partially disorder-specific.

In order to improve our knowledge of these problems we started a large-scale study program a few years ago. Our organizing criterion was to contemporaneously characterize patients according to a number of variables generally considered relevant for research on affective disorders. This was considered necessary for being able to apply computerized multivariate techniques that allow one to weigh the power and additive or interactive nature of associations, the prerequisite for acquiring an integrated picture of the inherent characteristics that are specific to affective disorders with ventricular enlargement. This chapter is a current, updated version of some previous works reported on this ongoing research program (10, 12, 13, 14, 15, 16, 17).

MATERIALS AND METHODS

Patients

The series of analyses to follow were performed on data from both in and out-patients followed at the Institute of Psychiatry of the University of Milan who had consented to undergo a non-contrast CT scan, had a previous history of at least three affective episodes and fulfilled the criteria for both Primary (18) and Major (19) Affective Disorder. All the patients also satisfied the DSM-III R criteria for Bipolar Disorder or Recurrent Major Depression (20). None of the patients had a documentable medical or neurological illness that might have caused ventricular enlargement or affective symptoms. A history of drug or heavy alcohol abuse, head trauma with loss of consciousness, seizures, electroconvulsive therapy or steroid intake during the 3 months prior to the CT were supplementary exclusion criteria. On this basis it was possible to recruit 135 patients (55 males, 80 females) between 24 and 69 years (mean age 52.4+9.8). Bipolar Disorder and Major Recurrent Depression were equally represented by chance (68 and 67 cases, respectively).

Patients were included in the various subanalyses only when they also fully satisfied supplementary specific criteria concerning age at onset, presence or absence of suicidal behavior, of definite anxiety and of mood congruent delusions, positive or negative family history and good or poor responsiveness to long-term lithium treatment during at least three years of follow-up (specific details on the procedures followed are given in 10 and 21 for all the variables except family history; those referring to the latter are reported in 22).

During the course of the study 21 patients (8 men, 13 women) gave their permission for a second CT scan after at least 23 months from the first examination, thus allowing a follow-up of cerebral ventricular size.

Controls

The control group consisted of subjects who had undergone CT scans either voluntarily or as part of routine diagnostic procedures to exclude brain injury from minor accidental head trauma without loss of consciousness. Individuals with neurological or CT signs even if only suggestive of possible accident-related lesions were excluded. Those with a personal or family history of any psychiatric disorder or who did not satisfy even one of the general criteria established for making the patients eligible for CT examination also were excluded. On this basis a group of 66 sex and age-matched controls was recruited.

CT Procedures

The CT scans were performed with a 1010 EMI scanner and with a 9000 II GE scanner in a smaller number of cases. Ten to twelve slices were obtained for each patient and control subject. The cerebral and ventricular measurements were made with a manual planimetric grid method (see 7 for details) on the CT slice showing the lateral cerebral ventricles at their largest by two independent raters who were completely unaware of the subjects' identities and had previously showed good inter-rater reliability (Pearson r=.95). Ventricular size was expressed as the ventricular brain ratio (VBR), that is, the ratio between the area of the lateral ventricles and the brain area multiplied by 100 (23).

Ventricular enlargement was defined as a VBR value exceeding the mean of controls by more than two standard deviations. Since VBR correlated directly with age in the control group, differentiated thresholds for defining ventricular enlargement were identified. The VBR limits thus were 7.5 up to 39 years, 8.1 between 40 and 49, 8.7 between 50 and 59 and 11.2 over 60 years.

RESULTS

Cerebral ventricular size in affective patients and controls

The VBR values of the patient group ranged between 0.8 and 16.4, and those of the control group between 1.0 and 10.2. The difference between the means of the two groups (6.0+3.6 and 4.7+2.5, respectively) was 1.3 units (t=2.5; p < .01). Looking at the data according to the different thresholds established for the various age groups, 25 patients, that is, 18.5% of the entire sample, had ventricular enlargement. Despite the lack of any apparent effect of sex on VBR among the controls, ventricular enlargement among the males in the patient group was 3.2 higher than in the females.

Cerebral ventricular size in affective patients

Time-related Variables

Cerebral ventricular size has been studied as a function of three main time-related variables: the age of the patient at the time of CT examination, the age at disease onset and the duration of illness. The latter did not correlate with VBR values (r=0.028; p=NS).

On the contrary, VBR values significantly correlated with both patient age (r=.37; p < .02) at the time of entrance into the CT scan project and age at onset of illness (r=.27; p <.05). Given the obvious interdependence of these variables (r=.53; p < .01), supplementary ad hoc analyses were performed on subsamples of early and late onset patients of the same age groups and on early onset patients under and over 40 so as to partially obviate the putative effects of aging on VBR on the one hand and of age at onset on the other. Compared to early onset peers (n=23), late onset patients (n=16) between 45 and 55 years had both higher VBR values (8+4.1 vs 4.2+2.4; t=3.3; p < .002) and increased chances for ventricular enlargement (8/16 vs 3/23; chi square 6.36; p <.01). In contrast, there was some overlap of either the VBR values (3.9+2.7 vs 5.6+3.6; t=1.6; p < .11) or the frequencies of cases with ventricular enlargement (1/15 vs 14/64; chi square 1.82) in the early onset patients under (n=15) and over 40 (n=64). Further, VBR and age positively correlated each other (r=0.28; p <.023) in both the early onset and control subjects, but this was not the case in the late onset patients (r=0.07; p=NS).

Clinical Subtypes and Symptoms

Case histories of manic episodes, depressive episodes with definite anxious or mood congruent delusional features and of suicidal behavior were checked for possible association with particular VBR values.

Major recurrent depression and bipolar disorder behaved similarly. Patients with the former diagnosis (n=67) did not have a cerebral ventricular size (6.4+3.6) significantly (t=.9; p=NS) greater than patients (n=68) with the latter one (5.7+3.4) and both types had almost identical frequencies of ventricular enlargement (11/67 vs 14/68; chi square .39; p=NS).

A relevant association was found in the cases of anxious or mood congruent delusional depressions. As compared to non-anxious subjects (n=52), definitely anxious patients (n=37) either had larger ventricles (ANOVA: F=7.3; p < .01) or higher frequency of cases with ventricular enlargement (11/37 vs 7/52; chi square 3.54; p <.057). Similarly, when delusional patients (n=28) were compared to non-delusional ones (n=89), the former had higher VBR values (ANOVA: F=10.7; p < .01) and increased chances for ventricular enlargement (9/28 vs 8/89; chi square 9.19; p<.003).

Suicidal behavior did not have an influence on ventricular size. Although patients with a history of suicidal behavior had a mean VBR exceeding that of the non-suicidal patient group (6.8+3.2 vs 5.7+3.4) and a 25% excess of cases with VE was found in the suicidal group (8/32 vs 19/100), neither of these differences were significant (t=1.5; p=NS and chi square .53; p=NS).

Clinical Outcome

Clinical outcome during long-term lithium or lithium-antiepileptic treatment also were investigated but in an exploratory manner in the latter case.

Ventricular size in 48 long-term lithium treated patients was evaluated. The 24 patients who, after a 3-5 year follow-up, were classified as responders had lower VBR values than the 23 non-responders (4.5+3.6 vs 7.4+3.4; t=2.9; p <.006). Furthermore, when ventricular enlargement was considered, it appeared that 7 of the 8 cases with this neuromorphological abnormality were lithium non-responders, while the majority of patients with normal ventricular size, 23 out of 40 cases, was satisfactorialy stabilized by long term lithium therapy (chi square 5.4; p<.019).

The preliminary data collected on long term lithium-dipropylacetamide or lithium-carbamazepine treatment indicated that ventricular size is less important in qualifying the response to these co-therapies, at least as regards patients previously classified as poor lithium responders. Among the 14 patients who had also received anticonvulsants for 2-3 years, better mood stabilization had been achieved in 4 of the 7 cases with ventricular enlargement and in 5 of the 7 patients with normal ventricular size.

Family History of Mood Disorders

Both a family history restricted to bipolar or major depression and a broader one of mood disorders were checked for possible interdependence with cerebral ventricular size.

In the first case, ventricular enlargement occurred in 36.5% of the 63 patients with no major affective disorders in their families and in 10% of the 50 patients with at least one secondary case among their first degree relatives (chi square 10.5; p<.002).

When the definition of positive family history was broadened to include all the mood disorders, the findings not only showed the same trend, but had their weight reinforced. Patients with a

positive family history of mood disorders (n=54) had both lower VBR values and a lower number of cases with ventricular enlargement when compared to patients (n=59) without affected relatives (4.99+3.2 vs 6.75+3.8; t=2.62; p < .01 and, respectively, 5/49 vs 23/36; chi square 13.36; p < .001).
The size of the families did not have any relevance as far as these results are concerned. The mean number of relatives in the families of patients with CT signs of central atrophy closely overlapped with that of those without this neuromorphological abnormality (5.46+2.1 vs 5.88+2.4; t=.867; p=NS).

Follow-up
The degree of stability of cerebral ventricular size in patients with affective disorders was evaluated by subjecting 21 patients to a second CT scan, at a mean time interval of 35.2+10.4 months from their first neuroradiological examination. VBR values had increased by as many as 1.5 units (5.8+3.8 vs 7.3+4.1; paired t=5.7;p<.0001). This largely exceeded the .3 units of the age-related physiological increase computed for controls of equal age.

CONCLUSIONS

The results of the various analyses presented call for both specific and general comments.
Specifically, we should first point out that when ventricular enlargement is present in affective patients it probably reflects processes that were operative at the earliest stages of the disorder or even before it. Consequently, the phenomenon cannot be viewed as a mere end-product generated by events such as drug treatment or repeated hospitalizations, events that are contingent on the evolution of the disorder. Substantiating this is the lack of association between cerebral ventricular size and duration of illness; however, the finding from the CT scan follow-up study, that showing a time-related increase of VBR values higher than that expected, clearly speaks for the neurodegenerative nature of ventricular enlargement, but does not completely exclude the possibility that treatment or other factors related to the course of the disorder might have some supplementary "enlarging" effect. Independent of the occurrence or non-occurrence of the latter, the lack of any apparent effect of aging on VBR values in the group of late-onset patients, that is, the group prototypically at increased risk for ventricular enlargement, appears compatible with the hypothesis that age-free neurodegenerative processes enlarging ventricular size in these patients are initially so redundant as to substantially overshadow the physiological effects of aging on VBR. Alternatively, we also could hypothesize that the progressive enlargement of the lateral cerebral ventricles by aging in a fair number of these patients is not only exponentially maximized but also variably anticipated that any correlation is cancelled.
Further, the fact that a late onset of the affective disorder has the power to influence the enlargement of the lateral cerebral ventricles _per se_ strongly supports the possibility that this neuromorphological abnormality phenotypically identifies a rather

definite subset of patients with inherently peculiar characteristics. Age at the onset of the disorder has been claimed to be a useful tool for clustering patients who are most similar from the symptomatologic, developmental, biochemical and familial viewpoints (13, 21).

Leading to similar conclusions for the same reasons are the findings that ventricular enlargement is more readily found in patients who scarcely benefit from long-term lithium treatment, who have, in the depressive phases, definite anxious or delusional features and few secondary cases of affective disorders among their first degree relatives.

In general, it must be emphasized that the accumulated data in this field already appear to represent an advance with respect to the past. In fact, our findings have always been concomitantly based on the same group of patients who have been characterized by a number of variables greater than those usually taken into consideration. Even though we cannot yet draw-up a unitary definitive picture because the necessary multivariate analyses are still to be undertaken, it still is probable that the variables so far independently associated with ventricular enlargement do represent different phenotypic manifestations of a homogeneous clinical condition. The association between a late disease onset and a depressive episode characterized by high levels of anxiety and delusions has internal coherence from the point at which it is validated by the first factor analyses (24) focused on depressive symptoms. Further, this picture closely resembling involutional melancholia (2), has long seemed familiar and acceptable to clinicians. On the other hand, the presumable neurodegenerative origin of ventricular enlargement further corroborates the relationship between involutional melancholia and affective disorders accompanied by this neuromorphological abnormality.

This technologically advanced revisit to involutional melancholia not only makes this disorder newly up-to-date, but even seems to extend the "coverage" pertaining to it. If we place the timing of the probable appearance of affective disorders with ventricular enlargement at around 40 years, it seems possible that the original descriptions of involutional melancholia ultimately represent not only late but particularly severe and disabling variants of a vaster group of affective disorders with CT scan signs of central atrophy.

Whether affective patients with ventricular enlargement suffer from a distinct disease entity or whether patients with and without this neuromorphological abnormality represent the two opposite extremes of a continuum of pathological states remains an unresolved question which requires specific future study, that is to say a return to the Kraepelin's doubts (25) about the independence or interdependence between involutional melancholia and the other affective disorders.

REFERENCES

1) Andreasen, NC (1988). Brain imaging: applications in psychiatry. Science, 239, 1381

2) Kraepelin, E (1921). "Manic Depressive Insanity and Paranoia". (Edinburgh: E.& S. Livingstone)

3) Weinberger DR, Wagner RL and Wyatt, RJ (1983). Neuropathological studies of schizophrenia: a selective review. Schizophrenia Bull, 9, 193

4) Roberts, GW and Crow, TJ (1987). The neuropathology of schizophrenia – a progress report. Brit Med Bull, 43, 599

5) Nagy, K (1963). Pneumoencephalographische befunde bei endogen Psychosen. Nervenarzt, 34, 543

6) Hounsfield, GM (1973). Computerized transverse scanning (tomography: part I. Description of system). Br J Radiology, 46, 1016

7) Sacchetti, E, Vita, A, Calzeroni, A et al. (1987). Neuromorphological correlates of schizophrenic disorder: focus on cerebral ventricular enlargement. In: Cazzullo, CL, Invernizzi, G, Sacchetti, E and Vita, A (eds.) "Etiopathogenetic Hypotheses of Schizophrenia: the Impact of Epidemiological, Biochemical and Morphological Studies". p. 67. (Lancaster: MTP)

8) Reveley, MA and Trimble, MR (1987). Application of imaging techniques. Brit Med Bull, 43, 616

9) Nasrallah, HA, Coffman, JA and Olson, SC (In press). CT and MRI findings in affective disorders: clinical and research implications. In Cazzullo, CL, Sacchetti, E, Conte, G et al. (eds.) "Morphology and Plasticity of the Central Nervous System". (Lancaster: Kluwer)

10) Sacchetti, E, Vita, A, Calzeroni, A et al. (1987). Ventricular enlargement: a marker for subtyping major affective disorders? In: Biziére, K, Garattini, S and Simon P (eds.) "Diagnosis and Treatment of Depression: Quo Vadis?". p.155.(Parigi: Medsi/McGraw-Hill)

11) Jacoby, RJ and Levi, R (1980). Computed tomography in the elderly. 3. Affective disorders. Brit J Psychiatry, 136, 270

12) Cazzullo, CL, Sacchetti, E, Vita, A et al. (1984). Cerebral ventricular size and age at onset in major depression. IRCS Med Sci, 12, 917

13) Sacchetti, E, Vita, A, Conte, G et al. (1986). Heterogeneity of major affective disorder and the outcome of long-term lithium treatment. IRCS Med Sci, 14, 407

14) Sacchetti, E, Vita, A, Conte, G et al. (1987). Cerebral ventricular size and clinical response to lithium prophylaxis in major affective disorder. Int J Neurosci, 32, 355

15) Vita, A, Sacchetti, E, Calzeroni, A et al. (1986). Computed tomography in psychiatric disorders. In: Reisner, T, Binder, H and Deisenhammer, E (eds.) "Advances in Neuroimaging". p.243. (Wien: Verlag der Wiener Medizinischen Akademie)

16) Vita, A, Sacchetti, E, Conte, G et al. (1987). The relationship between cerebral ventricular enlargement and anxiety symptomatology in major depression. In: Racagni, G and Smeraldi, E (eds.) "Anxious Depression: Assessment and Treatment". p.139. (New York: Raven Press)

17) Vita, A, Sacchetti, E and Cazzullo, CL (1988). A CT scan follow-up study of cerebral ventricular size in schizophrenia and major affective disorder. Schizophrenia Res, 1, 165

18) Feighner, JP, Robins, E, Guze, S et al. (1972). Diagnostic criteria for use in psychiatric research. Arch Gen Psychiatry, 26, 57

19) American Psychiatric Association, Committee on Nomenclature and Statistics (1980). "Diagnostic and Statistical Manual of Mental Disorders. 3rd Ed.". (Washington, DC: APA)

20) American Psychiatric Association, Committee on Nomenclature and Statistics (1987). "Diagnostic and Statistical Manual of Mental Disorders. 3rd Ed. Revised". (Washington, DC: APA)

21) Sacchetti, E, Conte, G, Calzeroni, A et al. (1987). The concept of anxious depression in relation to the biological heterogeneity of major affective disorders. In: Racagni, G and Smeraldi, E (eds.) "Anxious Depression: Assessment and Treatment". p.127. (New York: Raven Press)

22) Sacchetti, E, Calzeroni, A, Conte, G et al. (1988). Is ventricular enlargement a variable of interest for family studies in schizophrenia? In: Smeraldi, E and Bellodi, L (eds.) "A Genetic Perspective for Schizophrenic and Related Disorders". p. 217. (Milano: Edi Ermes)

23) Synek, V and Reuben, JR (1976). The ventricular brain ratio using planimetric measurement of EMI scans. Br J Radiology, 49, 233

24) Raskin, A, Schulterbrandt, JC, Boothe H et al.(1972). Some suggestions for selecting appropriate depression subgroups for biochemical studies. In: Williams, TA, Katz, MM and Shield, JA (eds.) "Recent Advances in the Psychobiology of the Depressive Illnesses". p.315. (Washington: US Government Printing Office)

25) Kraepelin, E (1913). "Psychiatrie III Band" (Leipzig: J.A. Barth)

Supported by CNR contract n. 87 00438.56 and Regione Lombardia grant n. 901.

8
Brain imaging with positron emission tomography in affective illness

M.S. Buschsbaum and J.C. Wu

INTRODUCTION

Brain imaging with positron emission tomography (PET) has the capability to survey functional activity throughout the brain and thus to bring together the disparate lines of neurochemical and behavioral approaches to affective illness. It has adequate resolution to view both individual gyri of the cortex and discrete portions of the basal ganglia, so important in depression and schizophrenia research. PET, using the mathematics of X-ray CT scanning to produce slice images of any chemical which is tagged with a radioisotope, opens an almost unlimited vista of metabolic studies. Positron-emitting atoms such as carbon 11 or fluorine 18 can be incorporated into sugar, amino acids, neurotransmitter precursors or psychoactive medications. Using a combination of such radiopharmaceuticals, one could potentially reveal the anatomic area made hypermetabolic by a delusion, the distribution of synaptic receptors for a treatment medication and the physiological effect of a drug with a repeat metabolic scan. This report reviews the initial studies of depression with PET and outlines some of the promise of the future.

METHODS

Typical deoxyglucose scan procedure

In a typical scan procedure with deoxyglucose in our laboratory, the patient arrives an hour before the scan. An intravenous line with a plastic cannula is inserted into a vein in each arm so that when the isotope arrives no time or isotope is lost. A psychological task is started just before isotope injection and must continue for 30-40 minutes. The tracer deoxyglucose is then injected in one arm and a series of 1 to 2cc blood

samples withdrawn from the other arm throughout the
uptake and scanning period for the measurement of
glucose and deoxyglucose. This arm is warmed with a hot
pad to increase arteriovenous shunting so that glucose
and 2-deoxyglucose concentration in samples approximates
that achieved with arterial sampling [1]. After 30-40
minutes, 70-90% of the deoxyglucose has been taken up
from the blood and the patient can be moved to the
scanner. The patient lies on a special bed, and his head
is placed into a holder to keep it still during the 3-15
minutes required to accumulate enough counts for an
image. An individually-fitted head holder made from
thermosetting plastic casting material to hold the
patient still during the scan and to allow the patient's
head to be returned to the same position for a second
scan on a later date. Now the coincidence gamma counts
are collected by the computer, from one or more rings of
crystals. The patient's bed may advance the patient
further into the ring to count additional slice
positions.

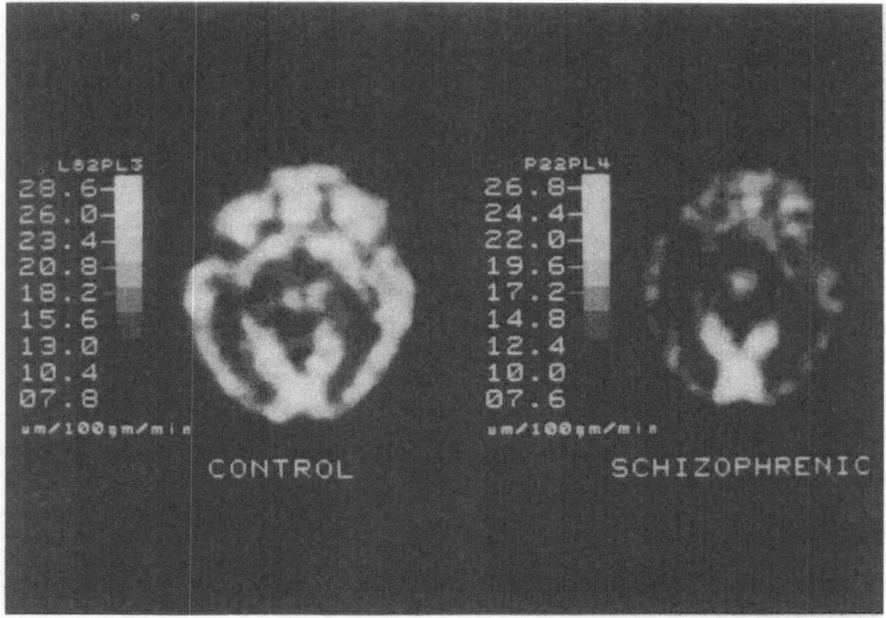

Figure 1. Positron emission tomography scans with
[18]F-2-deoxyglucose in normal control and patient with
schizophrenia. Scale is glucose metabolic rate in
micromoles glucose/100g brain/min. Note relatively
greater metabolic rate in frontal region in normal in
contrast to greater metabolic rate in occipital region
in patient with schizophrenia.

After acquisition of the count data, the computation of the final quantitative image begins (Figure 1). The image is typically smoothed to remove noise. A correction is made for the greater attenuation of radiation from central than peripheral tissue. Then, the data for deoxyglucose and glucose concentration in blood, together with their exact times, are united in a mathematical model of glucose metabolism [2,3] and the metabolic rate of the brain (in micromoles of glucose per 100 grams of brain tissue per minute) is calculated for each of the squares which make up the final PET picture. These numerical values are then transformed by the computer into the color images of brain activity. This transformation is linear over the range of physiologically normal tissue; thus, raw count and metabolic rate pictures have the same appearance except that the scale is now expressed in units of rate.

Imaging metabolic rate in affective disorder

Hypofrontality. Preliminary PET reports by Farkas et al [4] in one never-medicated schizophrenic showed a relative metabolic hypofunction of the frontal lobe. This was consistent with the pioneering studies of regional cerebral blood flow by Ingvar and Franzen [5]. Using intra-carotid injection of ^{133}Xe to assess regional flow, they observed that the ratio of frontal lobe to whole surface flow was reduced in patients with schizophrenia. Consistent cerebral blood flow changes in the frontal lobe have been found by a number of authors [6]. In the first published controlled series with PET, eight off-medication patients with schizophrenia and six normal controls [7] were studied. Patients rested in a darkened room with their eyes closed, to replicate the conditions of Ingvar and Franzen. The patients similarly had significantly lower frontal/whole slice glucose metabolic rate ratios (1.06) than normal controls (1.13). This was more strongly confirmed with linear trend analysis of variance indicating the presence of a front to back gradient in metabolic rate in both groups, stronger in normals than in patients with schizophrenia.

A second study (Figure 2 and 3) added a comparison group of patients with bipolar affective disorder. These patients were off medication a minimum of 14 days and a mean of 33.2 days, and were in the depressed phase of their illness. This study also found a reduced anteroposterior gradient in cerebral cortex in schizophrenia [8] as well as in the patients with bipolar affective disorder. These patients received brief electrical shocks to their right forearm during tracer uptake. This controlled task was chosen because of its reported tendency to increase cerebral blood flow [9] and because patients with bipolar affective disorder and schizophrenia had increased pain tolerance using

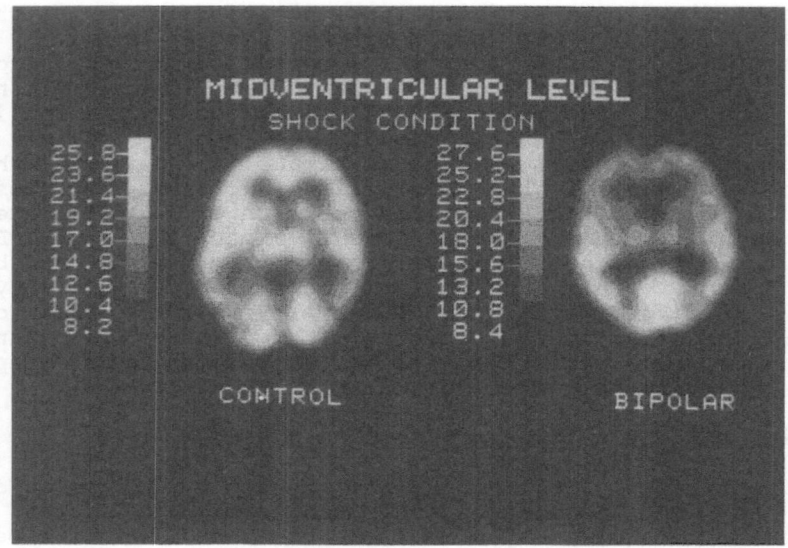

Figure 2. Normal control and patient with bipolar affective disorder off medication. Note marked decrease in relative metabolism in frontal region.

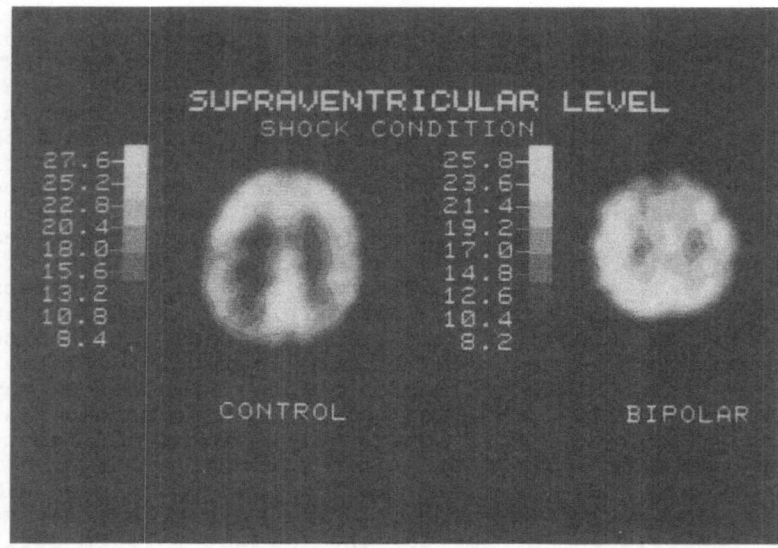

Figure 3.   At the level of the basal ganglia, patient with bipolar affective illness shows decrease in frontal lobe and in the basal ganglia.

exactly the same stimulation procedure [10,11]. Patients with bipolar affective disorder had significantly lower frontal to occipital cortex ratios of glucose metabolic rate (1.02) than normal controls (1.07). In this study, four patients with unipolar affective disorder actually had higher ratios of frontal to occipital cortex metabolic rate (1.17) than normals or bipolars [12]. Baxter et al [13] found not dissimilar results. While ratios were not reported, they noted that normal controls were 40% higher than bipolar depressed patients in the frontal cortex but only 29% higher in the occipital cortex. Their unipolar patients showed uniform differences (47% frontal, 46% occipital) with controls. However, we were unable to confirm the left>right hemispheric asymmetry they observed in bipolar patients.

Cortical metabolic rates across three slice levels in our sample were higher in patients with bipolar (25.3, SD=8.8) and unipolar disorder (28.6, SD=6.6) than in normal controls (19.7, SD=6.3). This is different from the finding of Baxter et al [13], who observed normal levels in unipolars and lower levels in bipolars. Differences between cortex and whole brain assessment, task during FDG uptake, duration of drug free interval may be important in these results.

The basal ganglia also showed a reduction in metabolic rate relative to the whole slice metabolic rate in patients with bipolar illness who were nearly all in the depressed phase (Fig. 3). This was significant for both the caudate and putamen as assessed by a stereotaxic region of interest method [12]. The effect was greatest in the dorsal caudate and putamen, as measured on the upper of two slices assessed (at 41% and 34% of head height above the canthomeatal line). Patients with unipolar depression had similar results. Baxter et al [13] also observed lower ratios of caudate to whole hemisphere metabolic rate, with depressed unipolar, manic bipolar, mixed bipolar and depressed bipolar groups all having mean ratios below the control groups. In this sample, only the unipolar group (n=11) reached statistical significance and the dorsal and ventral portions of the basal ganglia were not assessed.

In an analysis of the temporal lobe, affective and schizophrenic patients both showed increases, especially on the left side [14]. However, among patients with schizophrenia, greater increases were associated with more severe symptoms [15] whereas in the affective group the increases were seen among clinically improved patients. Patients actively depressed actually showed lower temporal lobe relative metabolic rates. Thus, while in the frontal lobes similarities between affectives and schizophrenics were observed, the pathophysiology of the temporal lobe in the two illnesses appears different.

Effects of antidepressants

Seventeen patients with major depressive disorder were studied by positron emission tomography with ^{18}F-deoxyglucose (FDG) after receiving placebo, imipramine or amoxapine for two days in a random assignment, double blind design. Patients performed the Continuous Performance Test, a visual vigilance test, during uptake of FDG. Nine slice images were obtained on the NeuroECAT scanner in the Visual Brain Imaging Center of the Department of Psychiatry at Irvine.

Data analysis focused on areas in the frontal lobes where we have previously observed decreases in glucose metabolic rate in patients with bipolar affective disorder and increases with unipolar affective disorder. We found significantly greater metabolic rates in frontal cortex were associated with amoxapine

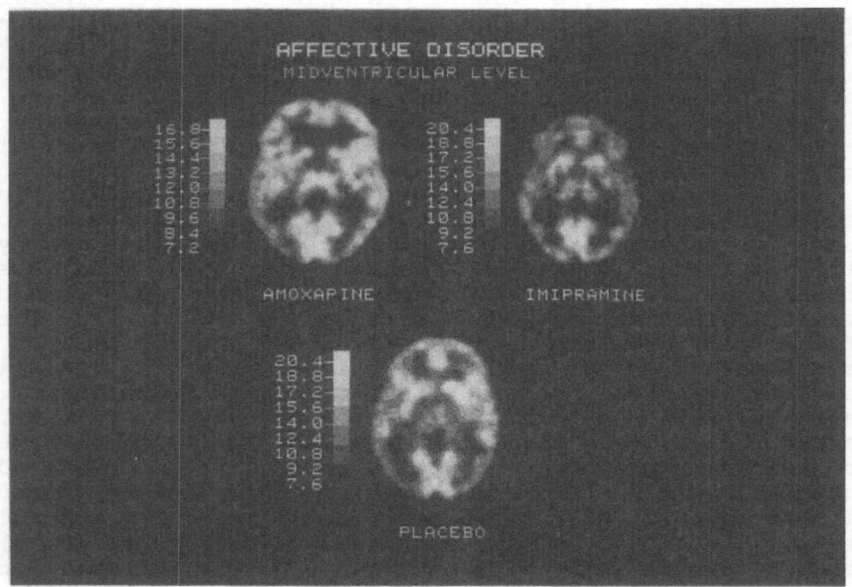

Figure 4. Patients with affective disorder after 48 hours of treatment with one of two antidepressants or placebo in a double blind acute trial. Note enhanced relative metabolic rate in frontal region of patient treated with amoxapine and decrease in patient treated with imipramine.

than placebo and lower metabolic rates with imipramine than placebo (Fig. 4). In occipital cortex, both drugs similarly elevated glucose metabolic rate.

Recently, it has been noted that amoxapine may have a more rapid onset of antidepressant action than other typical tricyclic antidepressants such as imipramine or amitriptyline. The possibility that this may be due to amoxapine's rapid ability to denumerize the serotonin S_2 receptor has been raised by some investigators [16]. The cerebral cortex contains high levels of serotonin S_2 receptors. Regional analysis also showed that serotonin S_2 receptors have their highest concentrations in the frontal cortex, with decreasing densities when more caudal regions of the CNS are approached [17]. Because we have previously observed that patients with affective illness have a reduction in glucose metabolic rate in frontal compared to occipital cortex [8], we hypothesized that we might observe frontal lobe metabolic rate changes with amoxapine earlier than with a standard tricyclic such as imipramine.

Sleep Deprivation

Sleep deprivation (SD) has been used as an experimental probe to infer what role sleep plays in behavioral and neurochemical activities and functions. Positron emission tomography (PET) scans were done to assess changes in local cerebral glucose utilization after SD. The hypothesis that caudate metabolism would be decreased in normal controls after SD was tested with PET scans. This hypothesis was based on previous studies suggesting changes in the dopaminergic system after SD.

Four normal controls (4 males, x = 25.3 ± 6 yr) were studied. The normal controls had no personal or family psychiatric history or significant medical illness. The subjects were scanned twice. The subjects were in a normal waking state on the first occasion and had slept the night before. The subjects were sleep-deprived for the entire night on the second occasion. The normal subjects felt moderately dysphoric after sleep deprivation.

The subjects performed the Continuous Performance Test, a visual vigilance test, during the thirty-minute uptake after injection of 4 to 5 mCi of 18-fluorodeoxy-glucose for both scans. Nine slice images were obtained on the CTI NeuroECAT IV scanner with an inplane resolution of 7.6 mm. Glucose metabolic rates for the scan were derived according to the model of Sokoloff.

Metabolic rates for the caudate are presented relative to the whole slice metabolism. Normal controls showed a significant decrease in relative caudate metabolic rates on the left side only using paired

t-tests   (p<.05,   1-tailed).   Relative   right   caudate
metabolic rates did not show significant change.
     Supersensitization and subsensitization in dopamine
system   have been found in total sleep or REM deprivation
using   behavioral,   neurochemical,   neuroendocrine,   and
receptor   studies   in   animals   and   in   man   [18,19,20].
Animal   studies   of   dopaminergic   agonists   effects on
caudate   metabolism   with   C14-labelled deoxyglucose have
found a positive relationship between dopamine agonist
potency   and   caudate   metabolism.   This study is
consistent with the possibility that the dopamine system
is   subsensitized   in   normal   controls after total sleep
deprivation.

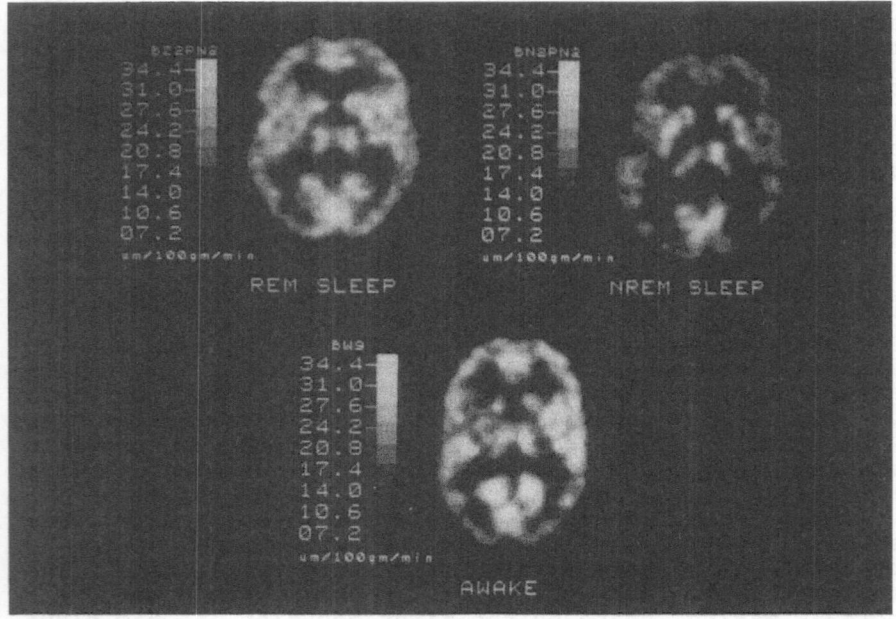

Figure   5.   PET scans at three standard levels before and
after sleep deprivation in one subject.   Images are at
same   metabolic rate scale.   Note regional differences in
metabolic change.

REFERENCES

1.   Phelps, ME, Huang, SC, Hoffman, EJ, Selin, C,
Sokoloff, L and Kuhl, DE (1979).   Tomographic
measurement of local cerebral glucose metabolic rate in
humans with (F-18)2-fluoro-2-deoxy-D-glucose: Validation of method.   Ann Neurol, 6, 371
2.   Sokoloff, L, Reivich, M, Kennedy, C, Des Rosiers,
MH, Patlak, CS, Pettigrew, KD, Sakurada, O and
Shinohara, M (1977).   The [^{14}C] Deoxyglucose method
for the measurement of local cerebral glucose utilization: Theory, procedure, and normal values in the
conscious and anesthetized albino rat.   J Neurochem,
28, 897
3.   Sokoloff, L (1982).   The radioactive deoxyglucose
method: Theory, procedure and applications for the
measurement of local glucose utilization in the central
nervous system.   In: Agranoff, BW and Aprison, MH (eds.)
"Advances in Neurochemistry". p. 1-82 (Plenum Publ.)
4.   Farkas, T, Wolf, AP and Cancro R (1980).   The
application of 18 F-2-deoxy-2-fluoro-D-glucose and
positron tomography in the study of psychiatric
conditions. The 12th Collegium Internationale Neuro-
Psychopharmacologicum Congress (Goteborg, Sweden)
5.   Ingvar, DH and Franzen, G (1974).   Distribution of
cerebral activity in chronic schizophrenia. The Lancet,
2, 1484
6.   Buchsbaum, MS and Haier, RJ (1987).   Functional and
anatomical brain imaging: Impact on schizophrenia
research.   Schiz Bull, 13, 113
7.   Buchsbaum, MS, Ingvar, DH, Kessler, R, Waters, RN,
Cappelletti, J, van Kammen, DP, King, AC, Johnson, JJ,
Manning, RG, Flynn, RM, Mann, LS, Bunney, Jr., WE and
Sokoloff, L (1982).   Cerebral glucography with positron
tomography in normals and in patients with
schizophrenia. Arch Gen Psych, 39, 251
8.   Buchsbaum, MS, DeLisi, LE, Holcomb, HH,
Cappelletti, J, King, AC, Johnson, J, Hazlett, E,
Dowling-Zimmerman, S, Post, RM, Morihisa, J, Carpenter,
W, Cohen, R, Pickar, D, Weinbeger, DR, Margolin, R and
Kessler, RM (1984).   Anteroposterior gradients in
cerebral glucose use in schizophrenia and affective
disorders. Arch Gen Psych, 41, 1159
9.   Ingvar, DH (1976).   Functional landscapes of the
dominant hemisphere.   Brain Research 107, 181
10.   Davis, GC, Buchsbaum, MS, Bunney, WE (1979a).
Research in endorphines and schizophrenia.   Schiz Bull,
5, 244
11.   Davis, GC, Buchsbaum, MS and Bunney, WE, Jr
(1979b).   Analgesia to painful stimuli in affective
illness.   Am J Psychiatry, 36, 1148
12.   Buchsbaum, MS, Wu, J, DeLisi, LE, Holcomb, H,
Kessler, R, Johnson, J, King, AC, Hazlett, E, Langston,
K and Post, RM (1986).   Frontal cortex and basal ganglia

metabolic rates assessed by positron emission tomography with ^{18}F-2-deoxyglucose in affective illness. J Affect Disorders, 10, 137

13.   Baxter, LR, Phelps, ME, Mazziotta, JC, Schwartz, JM, Gerner, RH, Selin, CE and Sumida, RM (1985). Cerebral metabolic rates for glucose in mood disorders. Arch Gen Psych, 42, 441

14.   Post, RM, DeLisi, LE, Holcomb, HH, Uhde, TW, Cohen, R and Buchsbaum, MS (1987). Glucose utilization in the temporal cortex of affectively ill patients: Positron emission tomography. Biol Psych, 22, 545

15.   DeLisi, LE, Buchsbaum, MS, Holcomb, HH, Langston, KC, King, AC, Kessler, R, Pickar, D, Carpenter, WT, Morihisa, JM, Margolin, R, Weinberger (in preparation). Increased temporal lobe glucose use in chronic schizophrenic patients

16.   Helmeste, DM and Tang, SW (1983). Unusual acute effects of antidepressants and neuroleptics on S_2-serotonergic receptors. Life Sci, 33, 2527

17.   Altar, CA, Kim, H and Marshall, JF (1985). Computer imaging and analysis of dopamine (D2) and serotonin (S2) binding sites in rat basal ganglia or neocortex labelled by [3H]spiroperiodol. J Pharmacol and Exper Therapeutics, 233, 527

18.   Zwicker, A and Calil, H (1986). The effects of REM sleep deprivation on striatal dopamine receptor sites. Pharm Biochem & Behav, 24, 809

19.   Lal, S et al (1981). Effect of sleep deprivation on dopamine receptor function in normal subjects. J Neural Transmission, 50, 39

20.   Gerner, R et al (1979). Biological and behavioral effects of one night's sleep deprivation in depressed patients and normals. J Psychat Res, 15, 21

9

The impact of starvation on brain morphology and function in eating disorders

W. Schreiber, C. Laver, K.M. Pirke, H.M. Emrich, G. Leinsinger, E.A. Moser and J.C. Krieg

INTRODUCTION

Brain imaging in psychiatry as well as in other medical fields aims at solving two fundamental problems:
1. Identification of structural alterations such as local or global atrophy.
2. Identification of functional alterations, for example of the regional cerebral blood flow or of metabolism.
Eating disorders, i.e. anorexia nervosa and bulimia nervosa, are characterized by continuous or intermittent fasting as a common means of losing weight. The question therefore arises whether the state of starvation can be held responsible for morphological and functional alterations found in both disorders. Pneumoencephalographic examinations, first carried out on dystrophic patients after years of war captivity, revealed a frequent enlargement of the anterior horns of the lateral ventricles, the anterior part of the cella media, the third ventricle and the external cerebrospinal fluid (CSF) spaces of the frontal area, presumably as a consequence of a brain edema due to protein depletion (1, 2). Comparable observations were made when investigating anorectic patients by means of post mortem autopsy (3), pneumoencephalography (4, 5) or cranial computed tomography (CCT) (6-15).
The aim of our investigations was to confirm these findings in a large sample of both anorectic (16, 17) and bulimic (18, 19) patients, with special reference to concomitant alterations of the adrenocortical and thyroid hormone status. In addition, we examined the regional cerebral blood flow (rCBF) by means of dynamic single photon emission CT (dSPECT), keeping the hypothesis in mind that the morphological brain alterations, reported to be associated with neuropsychological impairment in anorexia nervosa (10), are accompanied by abnormalities in functional CT diagnostics (20).

METHODS

All patients fulfilled the Research Diagnostic Criteria
of Feighner et al. (21) for anorexia nervosa and the
diagnostic criteria of Russell (22) for bulimia nervosa,
respectively, thereby meeting also the less stringent
DSM-III criteria (23). Ideal body weight (IBW) was registered according to the tables of the Metropolitan Life
Insurance Co. (24). In particular, 48 female and 2 male
anorectic inpatients (mean age $\pm$ SD : 21.5 $\pm$ 3.4; mean %
IBW $\pm$ SD : 69 $\pm$ 6%) and 50 female bulimic inpatients
(mean age $\pm$ SD : 22.5 $\pm$ 3.5; mean % IBW $\pm$ SD : 97 $\pm$ 10%),
respectively, were examined in our CCT studies; 48 female
and 2 male inpatients (mean age $\pm$ SD : 21.6 $\pm$ 3.6; mean %
IBW $\pm$ SD : 100 $\pm$ 9%) with a personality or adjustment
disorder but no organic neuropsychiatric disease or
substance dependence served as controls. Our rCBF study
comprised 12 female anorectic inpatients (mean age $\pm$ SD :
21.3 $\pm$ 3.0; mean % IBW $\pm$ SD : 73 $\pm$ 8%). Due to ethical
reasons, a sex- and age-matched control group of healthy
volunteers could not be established for this study.
Therefore, 5 female and 7 male subjects, whose rCBF had
been assessed to exclude a cerebrovascular disease and
who - on the grounds of a thorough examination - displayed no signs of a neuropsychiatric disorder, served as
controls (mean age $\pm$ SD : 31.0 $\pm$ 7.0; mean % IBW $\pm$ SD :
108 $\pm$ 15%). In addition, a CCT was performed on all
patients. CCT and dSPECT investigations as well as the
metabolite and hormone analyses were carried out as
described in detail by Krieg et al. (16-20), Büll et al.
(25,26) and Pirke et al. (27). Therefore, only essential
methodological features are presented here.

Enlargement of the external CSF spaces was assessed
by measuring the width of the cortical sulci, of the
anterior part of the interhemispheric fissure and of the
insular cisterns. A width of 3 mm and above shown by at
least six of these brain structures was defined as abnormal; in detail, a width of 3 or 4 mm was defined as a
"slight", whereas a width of more than 4 mm was defined
as a "marked" degree of sulcal widening.

A CCT scan through the area of the cella media of
the lateral ventricles, cut parallel to the glabella-
inion line (GIL), was chosen for the evaluation of ventricular size. Ventricular brain ratio (VBR) was assessed
by framing this region by a rectangle. Within this area
the ventricle province (density: -4 to +22 Hounsfield
units) and the total brain (density: -10 to +100 Hounsfield units) were measured in square centimeters by a
computer-aided program using a "density mask". The corresponding VBR values were calculated by dividing the
ventricular cross-sectional area by the brain cross-
sectional area and multiplying the result by 100.

rCBF was measured by the Xenon-133 inhalation method. Three axial slices (approximately 2 cm in thickness

with their centers 4 cm apart) 2 cm, 6 cm and 10 cm above the cantho-meatal line (CML) were examined, termed as slice 1, 2 and 3, respectively. For our investigations, only the area flow rates of slice 2 through the region of the lateral ventricles and of slice 3 through the convexity of the cortex were measured. These areas correspond approximately to the regions where the size of the ventricles and of the external CSF spaces, respectively, had been assessed by means of a CCT. The evaluation of the single flow maps was done by introducing 12 regions of interest (ROI) per slice, i.e. 6 ROIs per half-slice. These regions were delineated by a computerized program excluding a central band extending 5 pixels (i.e. about 1.7 cm) on each side of the midline. rCBF was defined as the average flow within the respective areas.

RESULTS

In our CCT studies we analyzed the scans of 50 anorectic inpatients. 41 (82%) of these anorectic patients displayed a "slight" (n = 23; 46%) or "marked" (n = 18; 36%) enlargement of the external CSF spaces, whereas only eight controls (16%) showed a corresponding sulcal widening (χ^2-test; $p < 0.001$). The patients with sulcal widening had a significantly lower body weight than the patients without such an enlargement (Mann-Whitney U-test; $p < 0.005$). On the metabolic and endocrine levels, respectively, patients with enlarged external CSF spaces displayed significantly higher plasma cortisol concentrations than patients without these morphological alterations (Mann-Whitney U-test; $p < 0.025$). This, however, was not the case in regard to other parameters of starvation such as the plasma levels of triiodothyronine (T3) and ß-hydroxybutyric acid (ß-HBA).

In 25 inpatients CCT controls could be performed both on admission and at discharge. At the second examination, which, on an average, took place 78 days after admission, anorectic patients had a mean body weight of 87% of their IBW; at this time, all parameters of starvation showed a significant trend towards normalization (t-test; $p < 0.05$ each). In approximately one fourth of these reexamined patients, the previous enlargement of the external CSF spaces had either disappeared or diminished parallel to weight gain.

Concerning ventricular dilatation, the anorectic patients displayed significantly higher VBR values than the controls (t-test; $p < 0.001$); in detail, 35 (70%) of the anorectic patients had VBR values that exceeded the highest VBR value within the control group. In accordance with the enlargement of the external CSF spaces, there was a significant inverse correlation between the VBR values and body weight (% IBW; Pearson product-moment correlation (r_p), $p < 0.05$). No correlation could be found between ventricular dilatation and the duration of anorexia nervosa. With respect to related metabolic and hormonal alterations, there was a close negative correlation between T3 values and VBR values (r_p, $p < 0.01$). In contrast to this, no significant association could be found concerning the VBR values and the other starvation parameters.

Similar to the partial reversibility of the enlargement of the external CSF spaces after weight gain, we also found a significant reduction of the mean VBR value (t-test; $p < 0.001$). In some cases, however, the morphological brain alterations remained unchanged even one year after weight normalization. Therefore two questions arose: First, whether the enlargement of the CSF spaces is exclusively related to the low body weight; and second, whether the reversibility of these morphological brain alterations depends merely on weight gain.

We therefore investigated the CCT scans of 50 normal weight bulimic patients. We chose this group of patients as it could be shown that despite their normal body weight many bulimics also display metabolic and hormonal signs of starvation, since they repeatedly attempt to lose weight by strict dieting (19, 27). Concerning external CSF spaces, an enlargement could be observed in 18 (36%) of the 50 bulimic patients. In particular, 15 (30%) subjects displayed a "slight" and 3 (6%) a "marked" degree of sulcal widening. The bulimic patients thus held a mid-position between anorectic patients and controls: They displayed enlarged CSF spaces significantly less often than the anorectic patients, but significantly more often than the controls (x^2-test; $p < 0.001$ and $p < 0.05$, respectively). In contrast to our observations in anorectic patients, there was no significant link between sulcal widening and body weight (% IBW); in addition, no correlation between sulcal widening and any of the starvation parameters could be found.

With regard to the VBR values, the bulimic patients displayed significantly higher VBR values than the healthy controls, but significantly lower VBR values than the anorectic patients (t-test; p 0.001 each); in detail, 22 (44%) of the bulimic patients had VBR values exceeding the highest VBR value found in the control group. No correlation between body weight and VBR values could be registered. Concerning metabolic and endocrine parameters, the patients with bulimia nervosa, like the anorectic patients, showed a significant inverse correlation between VBR values and T3 in plasma (r_p, $p < 0.05$). No association, however, could be found between the size of the ventricles and plasma levels of ß-HBA and cortisol, respectively.

In addition to these basic CCT findings, we investigated the regional cerebral blood flow in anorectic patients on admission and at discharge in a controlled study. There was a significant inverse correlation between the flow rates of slice 2 and VBR (Spearman rank-correlation (r_s), p > 0.05) and between slice 3 and the degree of sulcal widening (r_s, p < 0.05). Probably due to the high interindividual differences of the single flow rates, no significant differences between anorectic patients and controls could be registered concerning rCBF values in both slices. On the other hand, there was a strong correlation between the rCBF values and the variables % IBW, VBR and degree of sulcal widening (r_s, p < 0.05, each).

Dichotomizing the anorectic patients according to their weight gain after behavioral therapy resulted in 6 patients with a weight gain above and 6 patients below 10% of their IBW. Both groups had displayed comparable flow rates on admission; however, at discharge significant differences could be observed: Patients displaying a weight gain of more than 10% IBW had significantly higher rCBF values than those with a weight gain below this cut-off point (Mann-Whitney U-test; p 0.05). It is of special interest that the increase in rCBF values was most pronounced in the central region representing the area in which ventricular dilatation had diminished.

DISCUSSION

Taking all these results into consideration, the brain alterations observed in anorectic and normal weight bulimic patients are apparently caused by the process of starvation, regardless whether this results in emaciation or merely in fluctuations of the body weight due to counterbalancing bulimic episodes.

If the starvation-induced metabolic and endocrine disturbances are in fact relevant for the pathogenesis of the brain alterations, the partial maintenance after weight gain could be explained by intermittent relapses into fasting behavior common in both eating disorders. Two main parameters of starvation, i.e. elevated plasma cortisol and low T3 levels, have to be focused upon within this context.

With regard to increased adrenocortical activity, there are reports in the literature that patients with Cushing's disease (7, 28) or patients undergoing a corticoid therapy (29, 30) or suffering from a major depression with a concomitant hypercortisolism (31), display enlarged CSF spaces. Thus elevated steroid plasma concentrations, influencing catabolism of proteins, water and electrolyte distribution and vascular permeability, could be thought to be involved in the pathogenesis of the structural changes seen in starving anorectic and normal weight bulimic patients.

As far as low T3 levels are concerned, no specific impact on the enlargement of external or internal CSF spaces is known - at least to our present knowledge. There are only very few studies concerning morphological brain alterations related to hypothyroidism/myxedema. Besides a frequent enlargement of the pituitary gland only quite unspecific and poorly defined alterations were found, mainly in the cerebellar region resulting in reversible states of ataxia (32, 33). With respect to psychotic features in myxedema, Jellinek (34) reported a cerebral atrophy (i.e. either ventricular enlargement or excess of air over the cortex) in 6 out of 8 patients examined by means of pneumoencephalography; psychiatric reactions in myxedema were quite unspecific, comprising paranoid and depressive as well as agitated or manic states. Johnstone et al. (35) were able to confirm these observations in part by a CCT study in a sample of manic-depressive and neurotic outpatients, which revealed a significant association between increased VBR values and hypothyroidism. There are also some similarities between the EEG recordings of myxedema (34, 36, 37) and anorexia nervosa patients (38, 39) who often display low voltage and a slow frequency pattern .

The low T3-syndrome in anorectic patients is only in part comparable to the full clinical manifestation of a hypothyroidism or myxedema; likewise, the alterations described above are quite unspecific and do not allow a clear-cut pathophysiological interpretation. Nevertheless, the strong and consistent correlation between ventricular dilatation and T3 plasma levels provides some evidence for such a link. In addition, the question arises whether enlargement of external and internal CSF spaces is indeed caused by one and the same pathogenetic mechanism: According to the literature (7, 28-31, 34, 35) and our own results, one could speculate that hypercortisolemia and low T3 plasma levels are both involved in the pathogenesis of ventricular dilatation and sulcal widening but display a different mode of action and influence on the size of the external and internal CSF spaces. These different influences could also provide an explanation for our observation that sulcal widening was often - but not always - associated with ventricular dilatation. Further research seems to be worthwhile in order to clarify the different roles these starvation parameters play on brain morphology and function.

REFERENCES

1.   Faust, C (1949). Hirnatrophie nach Hungerdystrophie. Nervenarzt, 23. Jahrg., Heft 11, 406
2.   Schulte, W (1953). Hirnorganische Dauerschäden nach schwerer Dystrophie. (München, Berlin: Urban & Schwarzenberg)
3.   Gagel, O (1953). Die Erkrankungen des vegetativen Systems. In: Bergman von G, Frey, W and Schwiegk, H (eds.) "Handbuch der Inneren Medizin", vol. 5. p. 885. (Berlin, Heidelberg, New York: Springer)

4. Geisler, E (1953). Zur Problematik der Magersucht. Psychiatr Neurol Med Psychol, 5, 227
5. Heidrich, R and Schmidt-Matthias, H (1961). Encephalographische Befunde bei Anorexia nervosa. Arch Psychiatr Nervenkr, 202, 183
6. Enzmann, DR and Lane, B (1977). Cranial computed tomography findings in anorexia nervosa. J Comput Assist Tomogr, 1, 410
7. Heinz, ER, Martinez, J and Haenggeli, A (1977). Reversibility of cerebral atrophy in anorexia nervosa and Cushing's syndrome. J Comput Assist Tomogr, 1, 415
8. Nussbaum, M, Shenker R, Marc J and Klein, M (1980). Cerebral atrophy in anorexia nervosa. Brief clinical and laboratory observations, 96, 867
9. Sein, P, Searson, S, Nicol, AR and Hall, K (1981). Anorexia nervosa and pseudo-atrophy of the brain. Br J Psychiatry, 139, 257
10. Kohlmeyer, K, Lehmkuhl, G and Poutska, F (1983). Computed tomography of anorexia nervosa. A J N R, 4, 437
11. Artmann, H, Grau, H, Adelmann, M and Schleiffer, R (1985). Reversible and non-reversible enlargement of cerebrospinal fluid spaces in anorexia nervosa. Neuroradiology, 27, 304
12. Lankenau, H, Swigar, ME, Bhimani, S, Luchins, D and Quinlan, DM (1985). Cranial CT scans in eating disorder patients and controls. Compr Psychiatry, 26, 136
13. Sauer, H, Hornstein, Ch and Kessler, Ch (1985). Irreversible Hirnatrophie bei Anorexia nervosa. Nervenarzt, 56, 691
14. Datlof, S, Coleman, PD, Forbes, GB and Kreipe, RE (1986). Ventricular dilation on CAT scans of patients with anorexia nervosa. Am J Psychiatry, 143, 96
15. Deniker, P, Susini, JR, Ruyer, F and Fredy, D (1986). Apport de la tomodensitométrie cérébrale dans l'anorexie mentale. Etude de 16 cas. L'Encéphale, XII, 63
16. Krieg, J-C, Backmund, H and Pirke, K-M (1986). Endocrine, metabolic, and brain morphological abnormalities in patients with eating disorders. Int J Eating Disord, 5, 999
17. Krieg, J-C, Pirke, K-M, Lauer, C and Backmund, H (in press). Endocrine, metabolic and cranial computed tomographic findings in anorexia nervosa. Biol Psychiatry
18. Krieg, J-C, Backmund, H and Pirke, K-M (1987). Cranial computed tomography findings in bulimia. Acta Psychiatr Scand, 75, 144
19. Krieg, J-C, Lauer, C and Pirke, K-M (acc. for publ.). Structural brain abnormalities in patients with bulimia nervosa. Psychiatry Res
20. Krieg, J-C, Lauer, C, Leinsinger, G, Pahl, J, Schreiber, W, Pirke, K-M and Moser, EA (in prep.). Regional cerebral blood flow (rCBF) in anorexia nervosa
21. Feighner, JP, Robins, E, Guze, SB, Woodruff, RA, Winokur, G and Munoz, R (1972). Diagnostic criteria for use in psychiatric research. Arch Gen Psychiatry, 26, 57

22. Russell, GFM (1979). Bulimia nervosa: An ominous variant of anorexia nervosa. Psychol Med, 9, 429
23. American Psychiatric Association (1980). Diagnostic and Statistical Manual of Mental Disorders, 3rd ed. (Washington, D.C.: American Psychiatric Association)
24. Metropolitan Life Insurance Co. (1959). Statistical Bulletin of the Metropolitan Life Insurance Co., 40, 1
25. Büll$_{133}$U, Moser, EA, Kirsch, CM and Schmiedek, P (1983). ^{133}Xe-DSPECT (Dynamische Single Photon Emissions CT). Fortschr. Röntgenstr., 139, 351
26. Buell, U, Moser, EA, Schmiedek, P, Leinsinger, G, Kreisig, T, Kirsch, CM and Einhäupl, K (1984). Dynamic SPECT with Xe-133: Regional cerebral blood flow in patients with unilateral cerebrovascular disease: concise communication. J Nucl Med, 25, 441
27. Pirke, K-M, Pahl, J, Schweiger, U and Warnhoff, M (1985). Metabolic and endocrine indices of starvation in bulimia: a comparison with anorexia nervosa. Psychiatry Res, 15, 33
28. Momose, KJ, Kjellberg, RN and Kliman, B (1971). High incidence of cortical atrophy of the cerebral and cerebellar hemispheres in Cushing's disease. Radiology, 99, 341
29. Bentson, J, Reza, M, Winter, J and Wilson, G (1978). Steroids and apparent cerebral atrophy on computed tomography scans. J Comput Assist Tomogr, 2, 16
30. Gordon, N (1980). Apparent cerebral atrophy in patients on treatment with steroids. Dev Med Child Neurol, 22, 502
31. Kellner, CH, Rubinow, DR, Gold, PW and Post, RM (1983). Relationship of cortisol hypersecretion to brain CT scan alterations in depressed patients. Psychiatry Res, 8, 191
32. Church, A and Peterson, F (1903). "Nervous and Mental Diseases", 4th ed., rev. p. 496. (Philadelphia, New York, London: W.B. Saunders)
33. Weller, RO, Swash, M, McLellan, DL and Scholtz, CL (1983). "Clinical Neuropathology". p. 237. (Berlin, Heidelberg, New York: Springer)
34. Jellinek, EH (1962). Fits, faints, coma, and dementia in myxoedema. The Lancet, 3, 1010
35. Johnstone, EC, Owens, DGC, Crow, TJ, Colter, N, Lawton, CA, Jagoe, R and Kreel, L (1986). Hypothyroidism as a correlate of lateral ventricular enlargement in manic-depressive and neurotic illness. Br J Psychiatry, 148, 317
36. Levy, LL (1958). The EEG in thyroid and parathyroid disease. Electroencephalogr Clin Neurophysiol, 10, 366
37. Bertrand, I, Delay, J and Guillain, J (1938). L'électro-encéphalogramme dans le myxoedème. Compt Rend Soc de Biol, 129, 395
38. Crisp, AH, Fenton, GW and Scotton, L (1968). A controlled study of the EEG in anorexia nervosa. Br J Psychiatry, 114, 1149

39. Neil, JF, Merlkangas, JR, Foster, FG, Merlkangas, KR, Spiker, DG and Kupfer, DJ (1980). Waking and all-night sleep EEG's in anorexia nervosa. Clin Electroencephalogr, 11, 9

SECTION 3:
THE IMPACT OF
NEUROPHYSIOLOGICAL
STUDIES IN PSYCHIATRY

10
Neurometric subtyping of depressive disorders

L.S. Prichep, E.R. John, T. Essig-Peppard and K. Alper

Many researchers have demonstrated electrophysiological abnormalities in patients with depression. EEG spectral features [1-6], amplitude, asymmetry, mean integrated amplitude [7-14], and coherence [15-19], as well as features of the evoked potential (EP) [10,20-23], have all been identified as important in describing depressive disorders. The vast differences between studies in both the methods of acquisition and analysis of data make it extremely difficult to summarize this large body of data.

We have previously demonstrated the utility of Neurometric features of the EEG in the differential classification of psychiatric patients with a variety of disorders [24-31]. Neurometrics uses a standardized procedure in which 19 channels of monopolar EEG and EP data are collected with on-line artifact rejection. Quantitative features are extracted, log transformed to obtain Gaussianity, age-regressed, and Z-transformed relative to population norms. Z-values or standard scores for these features (proportional to probabilities) are then used for all further analyses. The importance of each of these steps in enhancing the clinical utility of electrophysiological data has been discussed in detail elsewhere [27,32]. In studies of large populations of children and adults, significant deviations from normal Neurometric values were rarely found in normally functioning persons, independent of culture or ethnic backgrounds [33-35]. However, significant departures from the normal range occurred frequently among patients with psychiatric disorders. Further, patients with different disorders showed distinctive Neurometric profiles of electrophysiological dysfunction.

Using small subsets of Neurometric EEG variables, we have previously reported high discriminant accuracy (split-half replicated, 91%) in separating major depression from normal as well as from other psychiatric disorders [28,36]. The composite feature of coherence in the anterior leads and parieto-occipital coherence in the delta band revealed incoherence in the depressed patients as compared to normal. Absolute power in the fronto-temporal regions was increased in these patients and was significantly asymmetric between hemispheres. Power asymmetries were also seen in the central regions.

While these features are useful in discriminating between major depression patients and normals, we have also found clear differences between the two subtypes

95

which make up this group. Figure 1 demonstrates this heterogeneity for one Neurometric variable: relative power in the beta frequency band in the right bipolar temporal derivation, T_4T_6. It can be clearly seen that when the unipolar and bipolar patients are combined into a "major depression" population, one would conclude that this variable does not differ from normal. However, dividing the population into subtypes reveals that the variable deviates in opposite directions from normal for bipolars (increased beta), and unipolars (decreased beta).

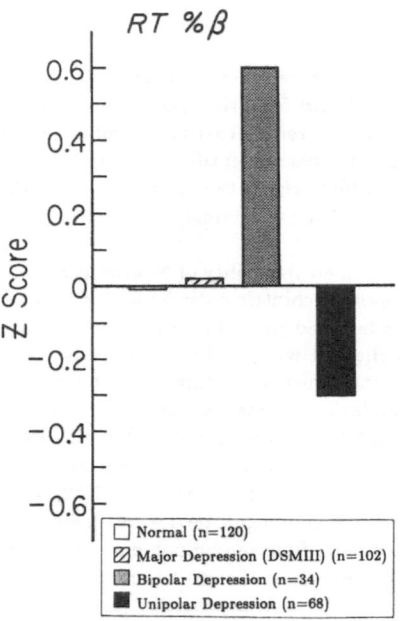

Figure 1: Bar graph of average Z-scores for relative power in the beta band in the bipolar temporal derivation (T_4T_6). The left bar is the average of the normal population (n=120). The second bar is the average for all the major depression patients (n=102). The depression group contained members of both the bipolar (n=34) and unipolar (n=68) subtypes. The averages for these two subtypes are shown in the third and fourth bars respectively.

In Figure 2, the group mean topographic head maps for Z-transformed monopolar relative power in the delta, theta, alpha and beta frequency bands are shown. Differences in Neurometric electrophysiological profiles can be seen between the bipolar and unipolar subtypes.

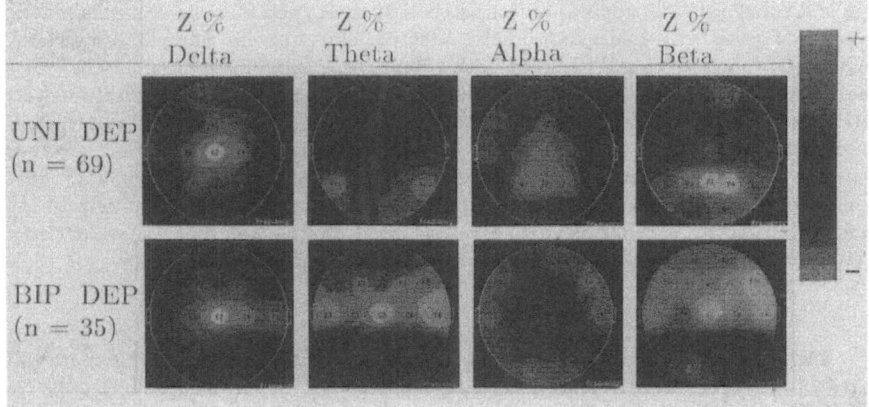

Figure 2: Average topographic head maps for Z-scores of relative power (%) in delta, theta, alpha and beta frequency bands, computed across groups of unipolar and bipolar individuals. These maps represent the mean relative power difference between each group and normative values, expressed in standard deviations of the reference (normal) group not shown in the figure. In the original figure these maps are color coded to reflect the mean Z-score. The scale goes from +1.5 to -1.5. The significance of Z-scale values can be estimated by taking the square root of the sample size and the standard deviation of each group into account. In this black and white reproduction, white represents both extremes of the scale and black the center, or zero point.

We have previously demonstrated an average accuracy of 88% (split-half replicated) in discriminating between these subtypes [28]. This high accuracy was achieved in classifying patients who were voluntary admissions to an inpatient evaluation unit at the time of study. In the present study, we have increased our sample to include an additional group of major depressed outpatients, with the intention of extending the practical applicability of our findings. We have also explored the utility of new quantitative evoked potential descriptors not used in our earlier work.

SUBJECTS

A sample of 151 patients were included with a diagnosis of major depression based on DSM III. These patients were divided into unipolar (n=108) and bipolar (n=43) subtypes. The population consisted of 59 males and 92 females with a mean age of 54 years. All had symptomatic affective illness severe enough to interfere with daily functioning and included both voluntary admissions to an inpatient evaluation unit and outpatients [1]. Psychiatric, medical and family histories, mental status evaluations and physical examinations were obtained. A Hamilton Depression Scale greater than or equal to 15 was required for admission to the study. Our previous studies required a Hamilton Depression Score greater than 17. In addition, the median was approximately 25 previously, as compared to approximately 20 in the current study. For a minimum of seven days prior to Neurometric evaluation, patients discontinued all medications except for antihypertensive therapy.

METHOD

Each patient received a Neurometric evaluation consisting of one minute of artifact-free eyes closed resting EEG and a battery of four evoked potential (EP) conditions. Monopolar recordings were obtained using the 19 electrodes of the 10/20 system [37], referred to linked earlobes. One hundred artifact-free trials were averaged in each EP condition. The EP conditions included:

1. Auditory EP's:

   (a) Random click: binaural clicks with a random interstimulus interval varying from 0.75 to 1.5 seconds.

   (b) Regular click: binaural clicks presented at a fixed interstimulus interval of 1.0 second.

2. Visual EPs:

   (a) Blank flash: binocular flash, rear projected, at a rate of one per second.

   (b) Grid flash: binocular 7 line/inch grid rear projected, at a rate of one per second.

---

[1]Major depression patients were evaluated as part of collaborative studies conducted with Dr. A. Lieber (Dept. Neuroscience, St. Francis Hosp. Florida); Dr. A. Georgotas (Depression Studies Program, New York Univ. Medical Center, Grant #MH35196), Dr. F. Mas (Dept. Psychiatry, New York Univ. Medical Center), as well as referrals to the Neurometric Evaluation Service, Dept. of Psychiatry, New York Univ. Medical Center

DATA ANALYSIS

A. EEG feature extraction:

The feature extraction methods have been described in detail previously [38]. Univariate and multivariate features were computed for absolute and relative power, coherence and asymmetry in the four frequency bands for the 19 monopolar derivations, as well as for eight bipolar derivations.

Since Z-scores express the deviation of the disparate Neurometric features from the predicted normative values in the common metric of relative probability, multivariate or composite features can be computed. Multivariates of two sorts were computed: (1) within each derivation across frequency bands for absolute power, relative power, coherence, or asymmetry, and (2) across derivations in the anterior or posterior regions of each hemisphere, across the whole left and right hemispheres and across the whole brain for every feature. Correction for intercorrelations among the features combined in each composite was accomplished by computing the Mahalanobis distance across the set of features. By procedures analogous to those used for univariate features, normative data were used to permit Z-transformation of these new composite features [38].

B. Factors for evoked potentials:

The factor structure of averaged visual and auditory evoked potentials in each of the 19 electrode derivations has been determined and normed by our laboratories. This method permits the evoked potential recorded from each scalp position to be described as a weighted combination of seven factor waveshapes. The weighting coefficients, or factor scores, are different for each electrode and in each stimulus condition. Using the norms which we have developed, these factor scores are then Z-transformed. Root mean square ($|\Sigma Z^2|^{\frac{1}{2}}$) factor scores, residual variances and total EP power are similarly computed. Details of this analysis are given elsewhere [39].

For each individual evoked potential factor Z-scores were computed for each of the 19 monopolar derivations, separately for random click, regular click, blank flash and grid flash. Z-scores were also computed for root-mean square factor scores, residual variances, and total EP power.

C. Discriminant functions:

The main purpose of this study was to extend and optimize our earlier "classifier" functions, which identified subtypes within a depressed inpatient population. The statistical procedure used was multiple stepwise discriminant analysis [40], which defines mathematical classifier functions, the values of which should be different for members of different *a priori* defined groups. These functions are weighted combinations of some subset of variables, each of which makes some independent contribution to the overall discrimination.

All patient populations were randomly divided into a "training" set (initial discriminant) and a "test" set (independent replication). The number of variables entered into the initial discriminant was reduced using methods such as the results of t-tests for the significance of differences between the groups [41] and surveys of the salient features reported in the literature. A subject-to-variable ratio of approximately 10:1 was used.

RESULTS

A. Classification of unipolar and bipolar subtypes using Neurometric EEG features alone:

Table 1 shows the accuracy of the initial discriminant classification and the independent replication for the separation of unipolar and bipolar depressed patients using *only* Neurometric EEG features. In the initial discriminant, ninety-one percent of the unipolar and eighty-three percent of the bipolar patients were correctly classified. In the independent replication, a mean discriminant accuracy of 75.5 percent was achieved. Seven Neurometric variables were used in the discriminant analysis (a subject to variable ratio greater than 20:1).

TABLE 1: Discriminant classification of unipolar vs. bipolar depressed patients using Neurometric EEG variables.

| | | Classification (%) as | |
Actual Group	n	I	II
		Initial discriminant	
I Unipolar Depression	54	91	9
II Bipolar Depression	23	17	83
		Independent replication	
I Unipolar Depression	54	76	24
II Bipolar Depression	20	25	75

B. Evoked potential factor Z-scores of unipolar and bipolar subtypes:

Many of the distributions of factor Z-scores were significantly different from the normals for both unipolar and bipolar subtypes, presumably reflecting features characteristic of major depression *per se*. Only a subset of these distributions however, was significantly different between the two subtypes. A representative subset of these features were subsequently used in the multiple stepwise discriminant computation described in section C. These included features with especially high t-tests plus some selected from the literature.

Figure 3 illustrates the mean topographic maps for each of these EP features for random click, separately for the unipolar and bipolar subtypes. Marked differences as well as similarities can be seen. EP features of random click were chosen for this illustration because they were preferentially selected by the multiple stepwise discriminant procedure.

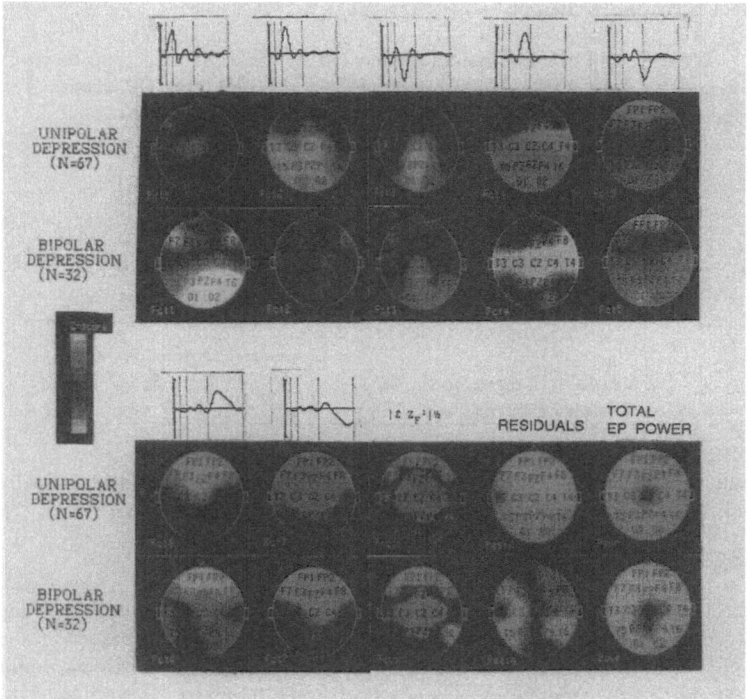

Figure 3: After Z-transformation of the factor scores for each factor for each electrode, group average topographic maps were constructed by averaging the factor Z-scores within the unipolar and bipolar groups separately for each factor and each electrode and interpolating the resulting values. The normative factor waveshapes are shown above the factor maps. In the original figure these maps are color-coded to reflect the mean Z-score. The scale goes from +0.75 to -0.75. The significance of Z-scale values can be estimated by taking the square root of the sample size and the standard deviations of each group into account. In this black and white reproduction, white represents both extremes of the scale and black the center, or zero point.

C. Classification of unipolar and bipolar subtypes using both Neurometric EEG and EP features:

Table 2 shows the accuracy of the initial discriminant classification and the independent replication for the separation of unipolar and bipolar depressed patients using *both* Neurometric EEG and EP variables. In the initial discriminant, ninety-eight percent of the unipolar and ninety-one percent of the bipolar patients were correctly classified. The mean discriminant accuracy in the independent replication was 79%. Seven EEG and seven EP variables were used in this discriminant analysis (a subject to variable ratio of better than 9:1).

TABLE 2: Discriminant classification of unipolar vs. bipolar depressed patients using Neurometric EEG variables and EP factor Z-scores.

Actual Group	n	Classification (%) as	
		I	II
Initial discriminant			
I Unipolar Depression	48	<u>98</u>	2
II Bipolar Depression	21	9	<u>91</u>
Independent replication			
I Unipolar Depression	45	<u>76</u>	24
II Bipolar Depression	17	18	<u>82</u>

It should be noted that differences in sample sizes in the various results herein reported reflect availability of data at the time of the computations.

DISCUSSION

This study demonstrates the utility of Neurometrics in the subtyping of depressive disorders. The separation of unipolar from bipolar patients based on a small subset of Neurometric EEG variables in this expanded population was accomplished with very much the same features as our previously reported discriminant [28]. In general, bipolar patients were characterized by diffuse increased beta activity, and decreased alpha activity (especially in the posterior lateral regions), while unipolars had more normal frequency distributions. Fromto-temporal absolute power asymmetries (with more power on the right side) also characterized bipolars. Unipolars had more posterior incoherence and more slow wave abnormalities (especially in absolute delta power in the right hemisphere). All of the Neurometric EEG features described above were entered and were useful in the combined EEG and EP discriminant. This confirms the importance of these features in describing these subtypes.

The results reported using EP factor Z-scores are preliminary but suggest increased classification accuracy when Neurometric EP features are added to EEG features. The composite features which so greatly increase the discriminant power of EEG variables have not yet been computed for the factor EP measure set. Thus, estimates of EP factor score contributions to discrimination based upon present results are probably quite conservative.

EP features from the random click condition were particularly useful in separating unipolar from bipolar depressed patients. Bipolar patients appear to differ from unipolars in the amount of variance accounted for by P_{200} in the EP to random clicks, especially in O_2 (also seen in O_1 and C_4), with bipolars significantly greater than unipolars. The overall EP power was greater in the unipolars than the bipolars in the anterior temporal and posterior regions (T_3, T_4, O_1, O_2, Pz). The residual variance not accountable by the normal factors was significantly greater in anterior temporal and posterior regions (T_3, T_4, O_1, O_2 and Pz), for random click in unipolars than bipolars. That is, significant variance outside the normal signal space occurred in EP's from these leads in response to random clicks in unipolar patients. The converse was true for the EP in response to regular click in derivation Pz for bipolars. Finally, unipolar patients showed significantly greater power in lead T_4 in response to a regular click.

Although initial discriminant accuracy was improved, we were somewhat disappointed by the fact that little was gained in discriminant accuracy of the independent replication when the EP factors were added to the discriminant. The overall discriminant accuracy of classification in a two group discriminant is reported for the 50% level. That is, if a discriminant classification score is greater than 0.50, the patient is considered to fall into the corresponding group. However, this tells us little about the *confidence* of an individual classification, which depends very much on the actual distributions of discriminant scores derived from the discriminant functions for the groups to be classified.

By considering classification accuracy curves at the $P < .10$, $P < .05$, and $P < .025$ confidence levels, the percentage of true classifications can be determined [32]. For the Neurometric EEG plus EP discriminant (Table 2), drastic improvements in classification accuracy curves were gained as compared to those for the discriminant based on Neurometric EEG features alone (Table 1). Classification accuracy at the $P < .50$, $P < .10$, $P < .05$, and $P < .025$ confidence levels for these two discriminants are shown in Table 3.

TABLE 3: Classification accuracy at different confidence levels using Neurometric QEEG alone and together with EP factor Z-scores

Classification	Classification Accuracy (%)							
	$P < 0.50$†		$P < 0.10$		$P < 0.05$		$P < 0.025$	
	EEG	EEG+EP	EEG	EEG+EP	EEG	EEG+EP	EEG	EEG+EP
Unipolar as Unipolar	84	87	56	66	20	65	6.5	38
Bipolar as Unipolar	21	14	10	10	5	5	2.5	2.5
Bipolar as Bipolar	79	86	42	82	31	55	28	35
Unipolar as Bipolar	16	13	10	10	5	5	2.5	2.5

†The maximum false positives in this case never go as high as 50%. Values shown are the mean of the original and independent replication groups.

Inspection of Table 3 shows that although the overall discriminant accuracy does not improve markedly at the $P < .50$ level, the confidence of classification drops off far more rapidly using EEG features alone than both EEG and EP features. The reader should note that "discriminant accuracies" in the literature are usually reported for the $P < .50$ confidence criterion, with no exploration of the shape of the classification accuracy versus confidence curve.

Shagass *et al* [7], O'Connor *et al* [16], Flor-Henry *et al* [17,18] and Ford *et al* [19] have previously reported the utility of coherence, absolute power and asymmetry characteristics of the EEG in computer classification of depressive patients. This study shows the further utility of these features in differentiating unipolar and bipolar subtypes in the depressed population. In addition, the observed differences in alpha and beta power between these subtypes may help to explain certain inconsistencies in previous reports.

Shagass *et al* [23] have similarly reported an increase in posterior early components of the auditory responses and others have found increased amplitudes in auditory responses recorded from a single vertex lead [42]. These anomalous findings in unlikely brain regions highlight the desirability of using an anatomically comprehensive set of electrode placements in electrophysiological studies of psychiatric patients. In fact, one way in which abnormalities are manifested in the depressed population is by paradoxically strong responses to stimuli in anatomically inappropriate brain regions.

Finally, confidence of classification curves, to which our attention was drawn by the pioneering work of Shagass and his collegues [7,23], appear to be indispensible for the proper evaluation of the potential clinical utility of discriminant functions which classify psychiatric patients.

Although the features derived from visual EP's did not enter the stepwise discriminant, this was due to the lower redundancy of auditory EP features in the combined EEG plus EP measure set, rather than to insensitivity of visual responses, which were also significant. We have not commented on earlier reports of augmenting/reducing for this reason.

This paper differs from most others in the literature in a number of important ways: Quantitative features extracted from the EEG and EP's were transformed to obtain Gaussianity, age-regressed, and Z-transformed against normative data; the intercorrelations between measures were removed from the EEG composite features; Z-transformed EP factor scores were used; independent split-half replications were performed for the discriminant analysis; and the confidence of classification was taken into account. Using this approach, the utility of electrophysiological measures of brain functions in the subtyping of depressive disorders has been confirmed and improved. Practical clinical utility may be forthcoming from use of both EEG and EP features to discriminate between patients who belong to unipolar and bipolar subtypes.

ACKNOWLEDGMENTS

The authors wish to acknowledge the contributions of Dr. Paul Easton and Dr. Jacob Friedman to this work.

This work was supported in part by Cadwell Laboratories Inc., Kennewick, Washington, U.S.A.

REFERENCES

1. Hurst LA, Mundy-Castle AC, Beerstecher DM (1954). The electroencephalogram in manic-depressive psychosis. *J Ment Sci*, 100, 220-240

2. Perris C (1966). A study of bipolar (manic-depressive) and unipolar recurrent depressive psychoses. *Acta Psychiat Scand Suppl,* 194, 118-152

3. Knott V, Waters B, Lapierre Y, Gary R (1985). Neurophysiological correlates of sibling pairs discordant for bipolar affective disorder. *Am J Psychiat*, 142, 248-250

4. Davis PA (1941). Electroencephalograms of manic-depressive patients. *Am J Psychiat*, 98, 430-433

5. Finley KH (1944). On the occurrences of rapid frequency potential changes in the human electroencephalogram. *Am J Psychiat*, 101, 194-200

6. Greenblatt M (1944). Age and electroencephalographic abnormality in neuropsychiatric patients. *Am J Psychiat*, 101, 82-90

7. Shagass C, Roemer RA, Straumanis JJ, Josiassen RC (1984). Psychiatric diagnostic discriminations with combinations of quantitative EEG variables. *Brit J Psychiat*, 144, 581-592

8. Endo S, Mori T, Kojima D, Akiyama M, Takagi H, Kuraoka Y, Kimura M (1987). Quantitative EEG and evoked potential study on hemispheric differences in Japanese depressive patients. In: Takahaski R, Flor-Henry P, Gruzelier J, Niwa S (eds) *Cerebral Dynamics, Laterality and Psychopathology*. Elsevier, Amsterdam, 85-92

9. d'Elia G, Perris C (1973). Cerebral functional dominance and depression. *Acta Psychiat Scand*, 49, 191-197

10. Perris C (1980). Central measures of depression. In: Van Praag H (ed) *Handbook of Biological Psychiatry*. Marcel Dekker, New York, pp 183-223

11. Flor-Henry P, Koles ZJ, Howarth BG, Burton L (1979). Neurophysiological studies of schizophrenia, mania and depression. In: Flor-Henry P, Gruzelier J (eds) *Hemispheric Asymmetries of Function in Psychopathology*, Elsevier, Amsterdam, pp 189-222

12. Kemali D, Vacca L, Marciano F, Nolfe G, Iorio G (1981). CEEG findings in schizophrenics, depressives, obsessives, heroin addicts and normals. *Adv Biol Psychiat*, 17-28

13. Schaffer CE, Davidson RJ, Saron C (1983). Frontal and parietal electroencephalogram asymmetry in depressed and nondepressed subjects. *Biol Psychiat*, 18, 753-762

14. von Knorring L, Perris C, Goldstein L, Kemali D, Monakhov K, Vacca L (1983). Intercorrelations between different computer-based measures of the EEG alpha amplitude and its variability over time and their validity in differentiating healthy volunteers from depressed patients. *Adv Biol Psychiat*, 13, 172-181

15. French CC, Beaumont JG (1984). A critical review of EEG coherence studies in hemisphere function. *Internat J Psychophysiol*, 1, 241-254

16. O'Connor KP, Shaw JC, Ongley CD (1979). The EEG and differential diagnosis in psychogeriatrics. *Br J Psychiatry*, 135, 156-162.

17. Flor-Henry P, Koles ZJ, Lind J (1987). Statistical EEG investigations of the endogenous psychoses: power and coherence. In: Takahaski R, Flor-Henry P, Niwa S (eds) *Cerebral Dynamics, Laterality and Psychopathology*, Elsevier, Amsterdam, 93-104

18. Flor-Henry P, Koles ZJ, Sussman PS (1983). Multivariate EEG analysis of the endogeneous psychoses. *Adv Biol Psychiat*, 13, 196-210

19. Ford MR, Goethe JW, Dekker DK (1986). EEG coherence and power in the discrimination of psychiatric disorders and medication effects. *Biol Psychiatry*, 21, 1175-1188

20. Buchsbaum M, Goodwin F, Murphy D (1973). Average evoked response in bipolar and unipolar affective disorders: relationship to sex, age of onset, and monoamine oxidase. *Biol Psychiatry*, 7, 199-212

21. Perris C, Eisenmann M (1980). Further studies of averaged evoked potentials in depressed patients with special reference to possible interhemispheric asymmetries. *Adv Biol Psychiat* 7, 199-212

22. Shagass C (1983). Evoked potentials in adult psychiatry. In: Hughs JR, Wilson WP (eds) *EEG and Evoked Potentials in Psychiatry and Behavioral Neurology*. Butterworth, Boston, 169-210

23. Shagass C, Roemer R, Straumanis J, Josiassen R (1985). Combinations of evoked potential amplitude measurements in relation to psychiatric diagnosis. *Biol Psych*, 20, 701-722

24. John ER (1977). *Functional Neuroscience, Vol II: Neurometrics, Clinical Applications of Quantitative Electrophysiology*. Lawrence Erlbaum Associates, Hillsdale, NJ

25. John ER, Karmel BZ, Corning WC, Easton P, Brown D, Ahn H, John M, Harmony T, Prichep L, Toro A, Gerson I, Bartlett F, Thatcher R, Kaye H, Valdes P, Schwartz E (1977). Neurometrics: Numerical taxonomy identifies different profiles of brain functions within groups of behaviorally similar people. *Science*, 196, 1393-1410

26. John ER, Prichep L, Ahn H, Easton P, Fridman J, Kaye H (1983). Neurometric evaluation of cognitive dysfunction and neurological disorders in children. *Prog Neurobiol*, 21, 239-290

27. John ER, Prichep LS, Fridman J, Easton P (1988). Neurometrics: Computer-assisted differential diagnosis of brain dysfunctions. *Science*, 239, 162-169

28. Prichep L (1987). Neurometric quantitative EEG features of depressive disorders. In: Takahaski R, Flor-Henry P, Niwa S (eds) *Cerebral Dynamics, Laterality and Psychopathology*, Elsevier, Amsterdam, 55-69

29. Prichep LS, Gomez-Mont F, John ER, Ferris S (1983). Neurometric electroencephalographic characteristics of dementia. In: Reisberg B (ed) *Alzheimer's Disease, The Standard Reference*. The Free Press, New York, 252-257

30. Prichep LS, John ER (1987). Neurometrics: Clinical applications. In: Remond A (ed) *Handbook of Electroencephalography and Clinical Neurophysiology, Vol III, Computer Analysis of the EEG and Other Neurophysiological Signals*. Elsevier, Amsterdam, 153-170

31. Prichep, LS, John, ER, Ahn, H, Kaye, H (1984). Neurometrics: Quantitative evlauation of brain dysfunction in children. In: Rutter, M (ed) *Developmental Neuropsychiatry*. Guilford Press, New York, 213-238

32. John ER, Prichep LS, Friedman J, Essig-Peppard T (1988). Neurometric classification of patients with different psychiatric disorders. In: Samson-Dollfus D (ed) *Statistics and Topography in Quantitative EEG*. Elsevier, Paris, 88-95

33. Ahn H, Prichep L, John ER, Baird H, Trepetin M, Kaye H (1980). Developmenal equations reflect brain dysfunction. *Science*, 210, 1259-1262

34. Alvarez A, Pascual R and Valdes P (1987). U.S. EEG developmental equations confirmed for Cuban schoolchildren. *Electroenceph Clin Neurophysiol*, 67, 330-332

35. John ER, Ahn H, Prichep L, Trepetin M, Brown D, Kaye H (1980). Developmental equations for the electroencephalogram. *Science*, 10, 1255-1258

36. Prichep LS, Lieber AL; John ER, Alper K, Gomez-Mont F, Essig-Peppard T, Flitter M (1986). Quantitative EEG in depressive disorders. In: Shagass C, Josiassen RC, Roemer RA (eds) *Electrical Brain Potentials and Psychopathology.* Elsevier, Amsterdam, 223-237

37. Jasper HH (1958). The ten-twenty electrode system of the international federation. *Electroenceph Clin Neurophysiol*, 10, 371-375

38. John ER, Prichep LS, Easton P (1987). Normative data banks and Neurometrics: Basic concepts, methods and results of norm constructions. In: Remond A (ed) *Handbook of Electroencephalography and Clinical Neurophysiology, Vol. III, Computer Analysis of the EEG and Other Neurophysiological Signals.* Elsevier, Amsterdam, 449-495

39. John ER, Prichep LS, Friedman J, Easton P (1988). Neurometric topographic mapping of EEG and evoked potential features: Application to clinical diagnosis and cognitive evaluation. In: Maurer K (ed), *Topographic Brain Mapping of EEG and Evoked Potential.* Springer-Verlag, Berlin, 90-117

40. *BMDP Statistical Software (P7M)*, (1987). University of California Press, Berkeley, California.

41. Weiner J, Dunn O (1966). Elimination of variates in linear discrimination problems. *Biometrics*, 22, 269-275

42. Friedman J, Meares R (1979). Cortical evoked potentials and severity of depression. *Am J Psychiatry*, 136, 1218-1220

11
Lateralization patterns of verbal stimuli processing in schizophrenia patients

S. Galderisi, A. Mucci, M. Maj and D. Kemali

INTRODUCTION

Research concerning lateralization patterns of verbal stimuli processing in schizophrenics has received a valuable contribution from studies using the divided visual field technique, consisting of very brief presentations of visual stimuli to the right or to the left visual field.

In fact, it has been consistently found that while normal subjects recognize verbal stimuli better and faster when tachistoscopically presented to the right visual field (RVF), schizophrenic patients generally fail to show such a visual field advantage [1, 2, 3].

Most of these investigations, however, have taken into account only behavioural measures of subjects' performance (reaction time and number of incorrectly identified stimuli). Therefore, little insight has been provided so far into mechanisms underlying the observed abnormalities: slower reaction time, for example, may be due to slowness in either response execution or stimulus evaluation.

In order to investigate the stage of brain information processing preceding the response execution, event-related potentials (ERPs), and in particular their late components, provide a very useful tool.

The late positive complex (LPC) is one of the most largely investigated ERP late components. It occurs about 300-900 msec post-stimulus and includes two different components: a peak, referred to as P300, and a slow positive activity, known as slow wave (SW).

LPC is elicited by task relevant, infrequent stimuli and is not dependent on the stimuli sensory modality. It represents a measure of stimulus evaluation time (that is of encoding, identification and categorization of stimuli) and is not influenced by response-related factors (planning, selection, control and execution of response) [4].

We believe that experiments combining this neurophysiological index with behavioural measures might provide a better understanding of lateralization pattern abnormalities in schizophrenia. Therefore, we decided to carry out the present investigation, in which ERPs, reaction time (RT) and number of incorrectly identified stimuli were simultaneously recorded in a group of DSM III diagnosed schizophrenics and in a closely matched control group, during a target detection task.

SUBJECTS AND METHODS

The selection criteria for subjects and the experimental procedure have been described in detail elsewhere [5, 6]. Briefly, patients were all male and their age range was 21-43 years. According to DSM III, 7 of them suffered from paranoid, 3 from undifferentiated, 3 from residual and one from disorganized schizophrenia. They had been drug-free for at least 15 days. The control group included 19 male healthy subjects, closely matched to patients with respect to age and educational level. All the subjects were right-handed, as assessed by the Oldfield´s questionnaire [7].

A visual target detection paradigm was used. Consonant pairs were projected on a screen by a projector provided with an electromechanical shutter, allowing a stimulus exposure time of 120 msec. Target stimuli, that is stimuli requiring the subject to press a button at their appearance on the screen, were same-name consonants; non-target stimuli (stimuli not requiring a response) were different-name consonants. Two experimental conditions were used: a central condition, in which stimuli were projected to the centre of the screen, and a lateral condition, in which the stimuli were presented to the right or to the left of a central fixation point.

The EEG signal was recorded from central and parietal leads (Cz, Pz, P3, P4) referred to linked earlobes. EOG was monitored from electrodes placed supraorbitally and at the lateral canthus of the right eye, in order to exclude trials contaminated with ocular artifacts.

Independent Principal Component Analysis (PCA) and varimax rotation were performed for ERPs relative to the central and the lateral condition. Analysis of variance was used to determine the effects of specific experimental variables (group, stimulus type, visual field, scalp location) on ERP components as reflected by PCA factors. Two-sample and matched t-tests were used to quantifye significant ANOVA effects and for statistical treatment of behavioural data.

During the same day of ERP recording, patients'
psychopathological state was evaluated by means of the
Comprehensive Psychopathological Rating Scale (CPRS [8])
and by the Scale for the Assessment of Negative Symptoms
(SANS) by Andreasen [9]. Pearson´s test was used to
investigate correlations between behavioural as well as
neurophysiological indices and scores on
psychopathological rating scales.

RESULTS

For centrally presented stimuli, the peak of the LPC was
reduced in schizophrenics as compared with controls, at
all scalp locations; the difference was significant at
the right parietal lead for target stimuli and at the
central midline lead for non-target (t=2.45, p<0.02 at
P4; t=2.69, p<0.02 at Cz) (Fig. 1).

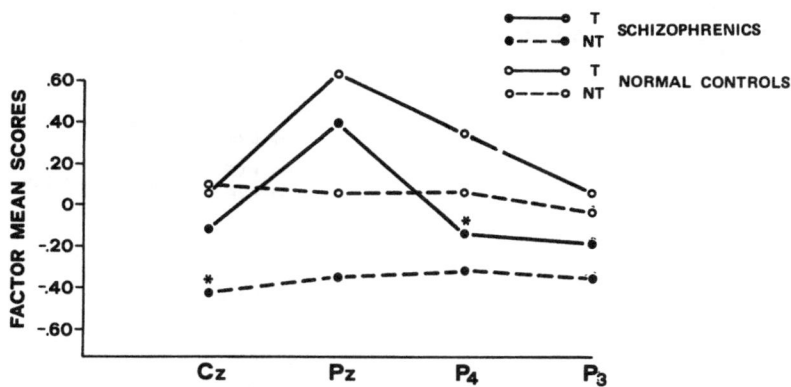

FIGURE 1   PCA factor 2 (LPC peak) mean scores for the
central condition. * Significant difference with
respect to normal controls, p<0.02

The LPC peak amplitude was inversely correlated
with SANS total score as well as with the scores on the
subscales Alogia, Anhedonia and Affective Flattening of
the same scale (table 1).

Table 1. Central condition: Correlation coefficient between LPC peak and SANS scores

| LEAD | S A N S   S C O R E S | | | |
	Affective Flattening	Alogia	Anhedonia	Total
Cz	-0.56*	-0.61*	-0.54*	-0.55*
Pz	-0.70***	-0.66**	-0.65**	-0.70***
P4	-0.53*	-0.67**	-0.70***	-0.65**
P3	-0.51	-0.60*	-0.56*	-0.48

* $p < 0.05$; ** $p < 0.02$; *** $p < 0.005$

Schizophrenics did not differ significantly from controls with respect to reaction time, number of errors and number of omissions.

For laterally presented target stimuli, the LPC peak amplitude in normal subjects was significantly higher for RVF than for LVF stimuli at Cz ($t=2.33$, $p<0.05$), P4 ($t=3$, $p<0.005$) and P3 ($t=2.1$, $p<0.05$); no visual field effect was observed in schizophrenics. Moreover, in the patient group, the amplitude of this component for RVF stimuli was markedly (although not significantly) reduced in comparison with controls (fig. 2).

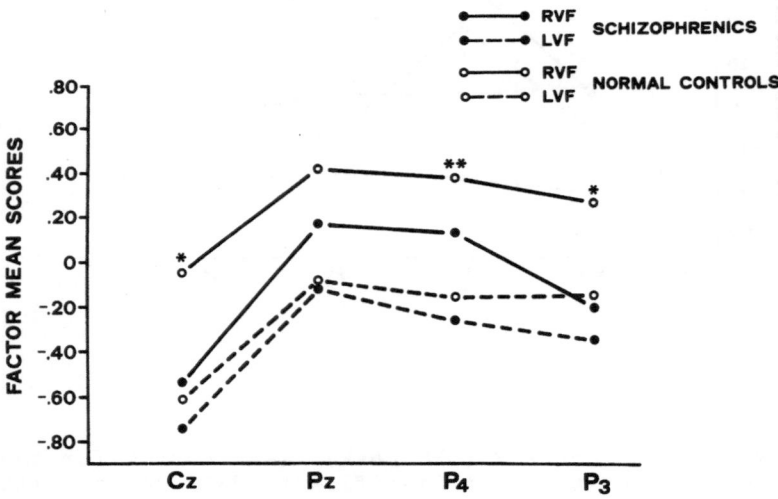

FIGURE 2. PCA factor 3 (LPC peak) mean scores for target stimuli in the lateral condition. Significant advantage of right vs. left visual field in normal controls, * $p<0.05$, ** $p<0.005$

As to the other component of the LPC, the slow wave, no visual field effect was observed in normal controls, while in the patient group the amplitude for LVF stimuli was larger than for RVF stimuli, at both midline leads (t=2.43, p<0.05 at Cz; t=3.06, p<0.01 at Pz)and at P3(t=2.69, p<0.02).

For RVF, in schizophrenic patients, the SW amplitude was lower than in controls at Cz (t=2.7, p<0.01), at P4 (t=2.41, p<0.02) and at P3 (t=2.46, p<0.02) (Fig. 3).

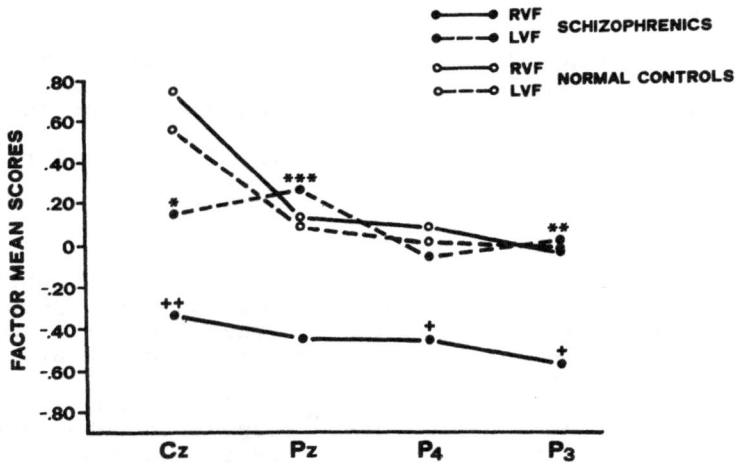

FIGURE 3. PCA factor 1 (Slow Wave) mean scores for target stimuli in the lateral condition. Significant advantage of left vs. right visual field in schizophrenics, * p<0.05, ** p<0.02, *** p<0.01. Significant difference between schizophrenics and controls in the RVF, + p<0.02, ++ p<0.01

Moreover, the SW amplitude for stimuli appearing in the LVF showed a significant positive correlation with CPRS global scores for positive symptoms (r=0.55, p<0.05).

No significant difference between patients and controls was observed with respect to the number of errors and of omissions. Reaction time results, on the contrary, showed that in normal controls stimuli presented to the right visual field were responded to significantly faster than stimuli presented to the left visual field, while in schizophrenic patients no visual field effect was observed.

Furthermore, patients were significantly slower than normals for RVF stimuli, whereas they did not significantly differ from normals for LVF stimuli (table 2).

Table 2. Lateral condition: Reaction time (msec, mean ± SD)

Subjects	Right Visual Field	Left Visual Field
Schizophrenics	704.2 ± 134.0*	707.4 ± 126.6
Normal controls	604.3 ± 85.6+	643.3 ± 97.2

* Significant difference between schizophrenics and controls, p<0.05
+ Significant advantage of right vs. left visual field in controls, p<0.05

DISCUSSION

Our data for centrally presented stimuli confirm previous findings of a reduction of the LPC peak in schizophrenia [10,11] and support the view that such reduction may be a neurophysiological correlate of negative schizophrenic symptomatology [12].

The lack of a significant difference between schizophrenics and controls with respect to behavioural indices (reaction time, number of errors and number of omissions) seems to rule out an effect of unspecific variables (such as lack of cooperation or distractibility) on the ERP findings.

The results obtained in the lateral condition suggest that in normal subjects the direct input of stimuli to the specialized hemisphere produces a faster response and a larger late positive complex peak. Since the LPC has been found to be free from response-related factors, its larger amplitude for RVF stimuli can be considered dependent on an easier encoding, identification and categorization of verbal stimuli when they are presented directly to the left hemisphere.

In schizophrenic patients, the lack of visual field effect on both RT and LPC peak consistently suggests a deficit of hemispheric specialization for verbal stimuli processing. Moreover, in these patients, the reverse visual field effect observed for the SW might be interpreted as an attempt at using right hemisphere linguistic abilities, which allow a correct, although less efficient, processing of verbal stimuli. On the basis of the correlational results concerning this component, such interpretation would hold true especially for patients with a prominent positive symptomatology.

In conclusion, our data, although preliminary, given the small size of the experimental sample, support the idea that ERP late components, in conjunction with behavioural measures, might represent a useful tool for studying abnormalities in information processing and hemisphere asymmetry of function in schizophrenic patients. Moreover, our ERP data suggest that positive and negative symptomatology may be associated with different patterns of abnormalities.

REFERENCES

1. Beaumont, JG and Dimond, SJ (1973). Brain disconnection and schizophrenia. Brit. J. Psychiat., 123, 661

2. Gur, RE (1978). Left hemisphere dysfunction and left hemisphere overactivation in schizophrenia. J. Abnorm. Psychol., 87, 226

3. Connolly, JF, Gruzelier, JH and Manchanda, R (1983). Electrocortical and perceptual asymmetries in schizophrenia. In Flor-Henry, P and Gruzelier, JH (eds.) "Laterality and Psychopathology". p.363. (Amsterdam: Elsevier)

4. Kutas, M and Hillyard, SA (1984). Event-related potentials in cognitive science. In Gazzaniga, MS (ed.) "Handbook of cognitive neuroscience". p.387. (New York: Plenum Press)

5. Kemali, D, Galderisi, S, Maj, M, Mucci, A, Cesarelli, M and D'Ambra, L (1987). Event-related potentials in schizophrenic patients: Clinical and neuropsychological correlates. Res. Comm. Psychol. Psychiat. & Behav., in press

6. Galderisi, S, Maj, M, Mucci, A, Monteleone, P and Kemali, D (1987). Lateralization patterns of verbal stimuli processing assessed by reaction time and event-related potentials in schizophrenic patients. Int. J. Psychophysiol., in press

7. Oldfield, RC (1971). The assessment and analysis of handedness: the Edinbourgh Inventory. Neuropsychol., 9, 97

8. Asberg, A, Montgomery, SA, Perris, C, Schalling, D and Sedvall, G (1978). A comprehensive psychopathological rating scale. Acta Psychiat. Scand., suppl. 271

9. Andreasen, NC (1981). Scale for the assessment of negative symptoms (SANS). (Iowa City: University of Iowa)

10.   Roth, W, Horvath, TB, Pfefferbaum, A, Kelley, AF, Berger, PA and Kopell, BS (1981). Auditory event-related potentials in schizophrenia and depression. Psychiat. Res., 4, 199

11.   Barrett, K, McCallum, WC and Pocock, PV (1986). Brain indicators of altered attention and information processing in schizophrenic patients. Brit. J. Psychiat., 148, 414

12.   Pritchard, WS (1986). Cognitive event-related potential correlates of schizophrenia. Psychol. Bull., 100, 43

12
EEG patterns in schizophrenia: a familial study

C. Colombo, O. Gambini, F. Macciardi, M. Locatelli, G. Calabrese and S. Scarone

INTRODUCTION

Visual assessment has shown various non specific EEG abnormalities such as poor alpha rhythm, increased beta activity and disorganized activity in schizophrenic patients as compared to healthy controls and subjects with other mental disorders. More recently, hypovariability of EEG activity, excess of beta activity and decreased alpha power have been demoonstrated by means of new sophisticated computerized techniques (1).

Subjects at risk for schizophrenia (i.e., children of schizophrenic mothers) show the same EEG patterns (increase of Beta activity) as those of chronic schizophrenic patients thus suggesting a possible genetic influence on these biological characteristics (1,2).

Further, a genetic control of ventricular enlargment has been suggested by morphological studies of schizophrenic twins (3). Increased abnormal eye movements have been found in relatives of schizophrenic patients (4). Data such as these again suggest a genetic determination of biological abnormalities in schizrenia.

In order to look at the same problem from a neurofunctional viewpoint we examined EEG characteristics in a sample of schizophrenic patients and their first degree relatives both under resting conditions (i.e. eyes closed and eyes open) and during the performance of two bimanual cognitive tasks.

MATERIAL AND METHODS

The sample consisted of 19 schizophrenics, 11 first degree relatives of these patients and 18 controls.

Two senior psychiatrists, who employed DSM III criteria (5), selected the subjects. The controls had no history of neurological or psychiatric illness. Before admission into the study they underwent physical and neurological examinations in order to exclude any possible disorders. All the subjects were right-handed, handedness having been assessed by means of a standardized questionnaire (6). The clinical and demographic characteristics of the study sample are reported in Table I.

Table I: DESCRIPTION OF THE STUDY SAMPLE

Controls :   N=18 (8M 10F);   x age 26.6 (9.6)
Patients :   N=11 (7M 4F);   x age 25.9 (5.8)
Relatives: N=19 (14M 5F);   x age 30.5 (13.9)

Schizophrenic type: Disorganized=4   Undifferentiated=5
                    Personality Disorders=2

EEG RECORDING
     The procedural and technical aspects of EEG recording have been described in previous papers (2). Briefly, subjects were seated in a soundproofed and electrically shielded recording room and the EEG recording was made continously throughout the four experimental sessions (i.e., Eyes Closed; Eyes Open; Simple Bimanual Spatial Task; Complex Bimanual Spatial Task).
     EEG measurements were taken using standard 10-20 electrode placements at P3 and P4, with reference at Cz and ground at Pz.
     The EEGs were recorded by means of an OTE BIOMEDICA 16 channel polygraph, the signal having been simultenously recorded on paper and analogic tape at a speed of 1 and 5/6 inch/sec with flutter compensation. It was analyzed by means of A-to-D conversion performed off-line by a Digital PDP 11 computer.
     The EEG recording was analog-filtered at 50 Hz, and stored on FM tape for off-line analysis. The signal was then digitalized at 250 Hz. Artifacts were removed after visual inspection and sequences of 2048 data points were subjected to Fast Fourier Transform (FTT) analysis. The log transformed power values for each of the following frequency bands were obtained over successive 8 second epochs: 1-4; 4-6; 6-8; 8-10; 10-13; 13-18; 18-30 Hz (i.e. delta, theta1, theta2, alpha1, alpha2, beta1, beta2) (7).

EXPERIMENTAL PROCEDURES
     The experimental protocol for the cognitive task was as

follows:
1) resting with eyes closed for 4 minutes; resting with eyes open
for 4 minutes;
b)   performance   of   a simple bimanual spatial   task   with   eyes
blindfolded   for   4   minutes (Simple   Task);   performance   of   a
complicated   bimanual   spatial task with eyes blindfolded   for   4
minutes (Complex Task).

The Simple Task required the subject to bimanually   examine
16   pairs of non-conventional wooden   geometric figures in   order
to identify the matching pair.

The Complex Task required the subject to bimanually examine
and   memorize   one of the 16 pairs of figures for   an   open-ended
period   (usually no more than 90 seconds).

Following this, the figures examined were put into a box with
the   others   and the subject then had to touch the objects   and
identify the pair of figures he had memorized.

Because   the   primary   aim   of the   cognitive   tasks   was   to
activate   the   parietal   area   (8),   we   did   not   evaluate   the
correctness   of the subjects'answers.   In order to motivate   them
and   solicit their full engagement we periodically assured   them
that   they   were   performing   well.

Although   we   did not apply standardized rating   scales   for
symptomatological   states,   our observations did not indicate any
hallucinatory behavior at the time of the test.

The   tactile   identification   of   non-conventional   geometric
figures seems to be of critical importance in schizophrenia   (9).
Research   from   our own   (10, 11) and   other   laboratories   (12)
support this hypothesis.

STATISTICAL ANALYSIS
Four   Discriminant   Analyses (13) were performed,   one   for
each different task (Eyes Open, Eyes Closed, Simple Task, Complex
Task).   The   log-transformed values of the relative power of   all
the frequency bands were the discriminant variables and diagnosis
(i.e.   control,   patient   and patient relative) was   the   grouping
variable.

RESULTS

Tables   II,III IV and V show the classification results   of
the Discriminant analysis:

Eyes   Closed Condition: the percent of grouped   cases   correctly
classified   was 77.8%;   45.5% of the cases were correctly grouped
as schizophrenics, 89.5% as relatives and 83.3% as controls.

Eyes Open Condition: the percent of grouped cases correctly classified was 66.7%; 83.3% of the cases were correctly grouped as schizophrenics, 63.2% as relatives and 45.5% as controls.

Simple Task Condition: the percent of grouped cases correctly classified was 68.7%; 45.5% of the cases were correctly grouped as schizophrenics, 79.0% as relatives and 72.2% as controls.

Complex Task Condition: the percent of grouped cases correctly classified was 70.8%; 63.6% of the cases were correctly grouped as schizophrenics, 68.4% as relatives and 77.8% as controls.

Table II: CLASSIFICATION RESULTS: EYES CLOSED

Actual group	No. of Cases	Predicted Group Membership		
		1	2	3
Group 1 CONTROLS	18	15 83.3%	1 5.6%	2 11.1%
Group 2 PATIENTS	11	3 27.3%	5 45.5%	3 27.3%
Group 3 RELATIVES	19	1 5.3%	1 5.3%	17 89.5%

Percent of "grouped" cases correctly classified: 77.08%

Table III: CLASSIFICATION RESULTS: EYES OPEN

Actual group	No. of Cases	Predicted Group Membership		
		1	2	3
Group 1 CONTROLS	18	15 83.3%	0 0	3 16.7%
Group 2 PATIENTS	11	2 18.2%	5 45.5%	4 36.4%
Group 3 RELATIVES	19	3 15.8%	4 21.1%	12 63.2%

Percent of "grouped" cases correctly classified: 66.67%

```
Table IV:  CLASSIFICATION RESULTS: SIMPLE COGNITIVE TASK
Actual group      No. of Cases      Predicted Group Membership
```

		No. of Cases	1	2	3
Group	1	18	13	1	4
CONTROLS			72.2%	5.6%	22.2%
Group	2	11	1	5	5
PATIENTS			9.1%	45.5%	45.5%
Group	3	19	3	1	15
RELATIVES			15.8%	5.3%	78.9%

Percent of "grouped" cases correctly classified: 68.75%

```
Table  V:  CLASSIFICATION RESULTS: COMPLEX COGNITIVE TASK
Actual group      No. of Cases      Predicted Group Membership
```

		No. of Cases	1	2	3
Group	1	18	14	1	3
CONTROLS			77.8%	5.6%	16.7%
Group	2	11	1	7	3
PATIENTS			9.1%	63.6%	27.3%
Group	3	19	4	2	13
RELATIVES			21.1%	10.5%	68.4%

Percent of "grouped" cases correctly classified: 70.83%

DISCUSSION

From a neurophysiological viewpoint,the four conditions tested in
this experiment (i.e., Eyes Closed Eyes Open, Simple Cognitive
Task, Complex Cognitive Task) were able to discriminate between
the three samples of subjects with a sensitivity of, at least,
63.2%.
In all the conditions tested the relatives were correctly
classified in their own group. This means that their EEG
characteristics are distinctive and different from those of
healthy controls. On the other hand, at least 75% of the
schizophrenics were classified in their own group or in the group
of relatives under all the experimental conditions. This
suggests some inherent susceptibiliy EEG trait in relatives of

schizophrenic individuals.

The Complex Cognitive Task appears to be the best condition for discriminating schizophrenic patients thus confirming the special aspects of EEG conditions during schizophrenics'mental activity. The inspection of the Univariate F-test performed on any single dependent variable in each condition (i.e. EC, EO, ST, CT) did not indicate either the specific involvemet of a single frequency band or a prevalence of one hemisphere in determining the discriminant function. Therefore these data, do not support the hypothesis of specific abnormalities in the beta band (14) or of a specific hemispheric malfunctioning in schizophrenia (15,16).

Acknowledgements: The Authors wish to thank Ms June Shmelzer La Rosa for her revision of the English text

This work was supported, in part, by the C.N.R. grant N. 87. 01626.04

REFERENCES

1) Itil, TM, Hsu, B, Saletu, B, and Mednick, SA (1974). Computer EEGs and auditory evoked potential investigation in children at high risk for schizophrenia. Am J Psychiat, 1731, 892 2) Colombo, C, Gambini, O, Macciardi, F, Bellodi, L, Sacchetti, E, Vita, A, Cattaneo, R and Scarone, S (1988). Alpha reactivity in schizophrenia and in schizophrenic spectrum disorders:demographic, clinical and hemispheric, assessment.Inter J Psychophysiol In press
3) Reveley, AM, Reveley, MA,and Murray, RM (1984). Cerebral ventricular enlargement in non-genetic schizophrenia: a controlled twin study. Br J Psychiat, 144, 89
4) Holzman, PS, Proctor, LR, Levy, DL, Yasillo, NJ, Maltzer, HY and Hurt, SW (1974). Eye tracking dysfunctions in schizophrenic patients and their relatives. Arch Gen Psychiat, 31, 143
5) American Psychiatric Association (1980). Diagnostic and statistical manual of mental disorders (3rd Ed), Washington D.C.
6) Razczkowski, D., Kalat, J.W. and Nebes, R. (1974). Reliability and validity of some handedness questionaire items. Neuropsychologia, 12,43–47.
7) Cooper, R, Osselton, JW and Shaw, JC (1980). EEG technology.Butterworths, London
8) Critchley, M (1953). The parietal lobes. Hafner, New York
9) Venables, PH (1980). Primary dysfunction and cortical lateralization in schizophrenia. In Koukkou, M, Lehman, D and Angst, J (Eds.) Functional states of the brain: their

determinants.p.243. (Amsterdam: Elsevier)

10) Scarone, S, Cazzullo, CL and Gambini, O (1987). Asymmetry of hemispheric functions in schizophrenia. Br J Psychiat, 151, 15

11) Gambini, O, Cazzullo, CL and Scarone, S (1986). Neurofunctional interpretation of abnormal responses to the Quality Extinction Test (QET) in schizophrneia. J Neurol Neurosurg Psychiat, 49, 997

12) Gruzelier, J, Seymour, K, Wilson, L, Jolley A and Hirsch, S (1988). Impairments on neuropsychychological tests on temporo-hyppocampal and fronto-hyppocampal functions and word fluency in remitting schizophrenia and affective disorders. Arch Gen Psychiat, 45, 623

13) SPSS user's guide (2nd Ed) (1986)(Chicago SPSS Inc.)

14) Koukkou, M (1982). EEG states of the brain, information processing and schizophrenic primary symptoms. Psychiat Res, 6, 235

15) Gruzelier, J (1984). Hemispheric imbalances in schizophrenia. Int J Psychophysiol, 1, 227

16) Flor-Henry, P, Fromm-Auch, D and Schopflocher,D (1983). Neuropsychological dimensions in psychopathology. In Flor-Henry, P and Gruzelier, J (Eds.). Laterality and Psychopathology. p.59. (Amsterdam: Elsevier Science Pub)

13

Some aspects of psychophysiological studies in psychiatry

J. Saarma, G. Morozow and N. Žharikov

In the activities of the central nervous system (CNS), composed from a number of subsystems, various levels can be differentiated. From the point of view of research it is practical to distinguish between four levels.

The basic, most elementary level of the life activity and of the adaptation of the CNS consists of biological processes of single nerve cells, including synapses between them.

Physico-chemical processes in the whole nervous tissue and in various subsystems of the CNS constitute another level.

Complex activities of various functional structures of the brain can be regarded as the next level of the CNS.

Behaviour as a highly integrated entity is the highest and most complicated level of CNS's activities. It consists of verbal, motor, vegetative and endocrine components.

By means of more and more sophisticated apparatus and more and more sensitive physical and chemical methods of investigation it has been possibl· in last decades to achieve marked results in studies into elementary levels and single functional structures of the CNS. It can be predicted even more marked success in this area in near future, e.g. by means of positron-emission tomography, qualitatively new electrophysiological, chemical and other research methods.

At the same time progress in studies into complex behaviour of normal and mentally disturbed man is comparatively moderate. In last years clinical investigators have had to be sufficient with classical descriptive psychopathological methods.By means of various general and specific rating-scales certain objectivity and standardization has been achieved in registration of clinical data.

In clinical studies into neurophysiological background of psychic disturbances and into bases of the action of contemporary treatment methods there is a real necessity to make use of more differentiated methods of investigation as a supplement to clinical observation.

First of all, psychometric methods known in experimental
psychology may serve as suitable tools in these investigations. Mental diseases are manifested first of all in
various disturbances of behaviour, i.e. of the integrated verbal-motor-vegetative complex reactions. Of course
it is important to study into and to know physical and
chemical basis of these disturbances, but this alone is
not enough. It is also necessary to know measurable and
comparable parameters of the behaviour itself. By means
of psychometric tests it is possible to gain more exact
parameters in this field. It is important to study into
and to know physico-chemical basis of the action of various therapeutical measures, but this alone is not en-
ough.Our aim in treating a patient is not to bring about
certain changes in the activities of some mediator in
his limbic system, but to attain certain changes in patient's behaviour. Psychometric tests allow to measure
these changes more exactly.
Many, mainly Soviet investigators have demonstrated,that
using basic concept of I. PAVLOV on the higher nervous
activity (HNA) to interprete the data of psychometric
investigations the neurophysiological background of behaviour and it's disturbances can be understood in a more
systematic way. Some investigators suppose, that in connection with marked successes of electrophysiological,
neurochemical and psychopharmacological investigations
Pavlovian concept of the HNA has lost it's meaning. This
opinion is not founded. It is well known that for investigation of different levels different methods must be
applied. A research method adequate for one certain level is not applicable to study into another level. Facts
and rules found out by means of one method in one structure cannot be universally applied in understanding activities of other levels. In other words: various research methods do not compete with but complement each
other, everyone of them having a specific field of application. Until now no better concept to systematically
interprete cortical activities in a complex way as Pavlovian one has been proposed.
Data collected by means of various methods summarized
and systematized deliver more complex knowledge about
activities of normal and disturbed CNS. In studies into
cortical processes, i.e. into HNA of normal and mentally
ill person it is also necessary to make use of different
psychometric methods. In this way it is possible to gain
more complex picture about disturbances and their dynamics in cortical activities,i.e. in elements of the complex behaviour of the person. Every psychometric test
characterizes a specific cortical function, so it's data
can be interpreted only regarding to this certain function. A number of psychometric tests applied in investigation of a person can deliver complex data, characterizing more generally the behaviour of the person.

As one possible complex Tartu Psychometric Tests Battery may be recommended (table 1).

Table 1. TARTU PSYCHOMETRIC TESTS BATTERY

METHOD OF INVESTIGATION	CORTICAL FUNCTION INVESTIGATED
OPERANT MEMORY TEST	Connecting activity of the second signalling system (ecomemory, operant memory)
LEARNING TEST	Connecting activity of the second signalling system (fixation)
ASSOCIATION TEST	Activity of old free conditional connections of the second signalling system
CALCULATION TEST	Activity of old automatized conditional connections of the second signalling system
CORRECTION TEST (with letters)	Cooperative activity of the first and second signalling systems
MOTOR REFLEX TEST	Activity of the first signalling system and it's elementary connections with the second signalling system
VEGETATIVE COMPONENT OF THE ORIENTING REFLEX	Activity of cortico-vegetative connections

In compiling the battery we tried to make sure, that: a) every test applied is simple and well understandable to every normal as well as disturbed person and b) data of every test is expressable in numbers. Results of investigations by means of our battery we interpreted in the universal Pavlovian concept on the HNA (table 2). Experiences with this battery of psychometric tests in a number of years testify to usefulness of results of these investigations both in theoretical understanding of neurophysiological basis of behavioural disturbances and in practical psychiatry. This is illustrated by following main results of our studies.

1. It has been possible to find out some specific features of the cortical activities in various mental diseases. These peculiarities can serve as complementary criteria in differential diagnostics of diseases. In schizophrenia most characteristic feature is a moderately intense transmarginal inhibition of cortical functions, mainly in the form of hypnotic phasic conditions with pulsating intensity. This phasic inhibition is found not only in cortical functions, but also in cortico-vegetative connections, as well as in central regulatory mechanisms of vegetative functions.

Table 2. PROCESSES OF THE HNA CHARACTERIZED BY
TARTU PSYCHOMETRIC TESTS BATTERY

PARAMETERS MEASURED BY THE TESTS APPLIED	PROCESSES OF THE HNA CHA-CHARACTERIZED
Latency periods, power and quality of reactions	EXCITATORY PROCESS
Adequacy and quality of re-actions	INTERNAL (DIFFERENTIAL) INHIBITORY PROCESS
Stability of latency periods of reactions (median devia-tions)	STABILITY OF NERVOUS PRO-CESSES
Stability of latency peri-ods, power and quality of reactions under various con-ditions of investigation	TENDENCY TOWARDS EXTERNAL INHIBITION OF CORTICAL ACTIVITIES
Action of signals in one signalling system upon reac-tions by the other signal-ling system	COOPERATION OF SIGNALLING SYSTEMS

In depressive states a more stable transmarginal inhibi-tion of cortical, especially effectory functions is cha-racteristic. This inhibition is connected with and obvi-ously induced by a focus of steady excitatory process in subcortical structures.

In neuroses the most characteristic disturbance in cor-tical activities is instability of internal (differenti-al) inhibitory processes as well as of excitatory pro-cess, especially in the second signalling system and in cortico-vegetative connections.

It has been also possible to find out differences in these disturbances of cortical activities in connection with duration as well as intensity of mental diseases.

2. The action of many therapeutical measures (insu-lin treatment, electroconvulsive treatment, ammonium treatment, psychotropic drugs - including a great num-ber of neuroleptics, antidepressants, tranquilizers and nootropic drugs) upon cortical activities has been ex-plored. Marked differences in the action of even similar treatment modalities have been found in various conditi-ons and in different patients. It has been demonstrated that data of systematic complex psychometric investiga-tions characterize quality of therapeutic shifts more deeply than clinical observation alone. In many cases experimental data have been helpful to recognize that the remission is only superficial and thus contributed to prevention of relapses.

3. On the basis of vast clinical experiences it has been possible to find out some specific features in the psychometric data complex, which can serve as comp-

lementary criteria for certain treatment measures. So these investigations contribute to a more rapid finding out the most adequate and effective therapeutic method for a certain patient.

On the basis of our accumulated experience we come to the conclusion, that in contemporary situation of marked efficacy of fundamental biological research into mental diseases psychometric investigations, especially when their results are interpreted in a systematical way on the basis of Pavlovian concept on the HNA continue to have a marked place among other methods and attitudes. They contribute essentially to a more complex understanding of behavioural disturbances and to a more effective realization of therapeutic possibilities in patients suffering from psychic diseases.

SECTION 4:
THE IMPACT OF GENETIC STUDIES IN PSYCHIATRY

14
Genetic marker studies in schizophrenia

L.E. DeLisi

INTRODUCTION:

A strong familial component to the etiology of schizophrenia was clearly recognized by Kraepelin when he
attributed as many as 70% of cases to a "defective
heredity factor" (Kraepelin, 1907 [1]). This was later
confirmed over the last century in systematically examined cohorts of patients and their family members,
monozygotic and dizygotic twin concordance studies, and
the more recent series of adoption investigations (reviewed by Gottesman and Shields, [2]). Little, however,
has yet been uncovered to establish a specific mechanism
for genetic transmission and its biological substrate(s).
One possibility is that this lack of progress is a
consequence of biological heterogeneity to the clinical
syndrome called schizophrenia. Alternatively, the presently defined clinical syndrome may be inadequate to
delineate specific biologic markers and the relevant
clinical boundaries to the biologic syndrome still un-
determined. The reverse of "biological heterogeneity"
to schizophrenia may be "clinical heterogeneity" to one
biological disease—the proposal that schizophrenia and
affective disorder are at opposite ends of a continuum
of psychosis [3]. In support of the latter are family
studies that show an increase in affective disorder in
the relatives of patients with schizophrenia [4,5] and
an increase of schizophrenia in relatives of patients
with bipolar affective disorder [4], implying a genetic
overlap between affective disorder and schizophrenia.
Schizoaffective disorder, the condition with symptoms
of both schizophrenia and mood change, poses an additional problem and supports the continuum hypothesis,
since no genetic boundary can be drawn between this
disorder, schizophrenia and affective illness [6,7].

It is therefore likely that biologic markers for the genetics of psychosis will cross clinical diagnostic boundaries. A marker associated with one of these diagnostic categories should be considered relevant to the others.

THE USE OF GENETIC MARKERS:

Genetic markers are direct indicators of genes or their products (i.e. DNA, RNA, or protein). The considerable variation in DNA base-pair sequences (e.g. polymorhisms) that normally occurs between individuals can be used to "mark" the inheritance of specific genes and whether they segregate with certain diseases within families. A positive association, in genetic terms, signifies a higher frequency of a specific marker in a sample of nonrelated patients compared with a control population, and suggests a relationship of that polymorphism with the illness etiology. A positive linkage study, however, makes use of the polymorphism only to trace the inheritance of a chromosomal region from parent to offspring. The polymorphism, itself, as well as the gene it represents, is not implied to be related to the cause of the illness.

CHROMOSOMAL ABERRATIONS IN PSYCHOTIC PATIENTS:

Microscopic examination of chromosomal structure in patients with psychosis, in general, have been negative, but have not been adequately performed, given the newer cytogenetic culture and banding techniques. Most of the chromosomal studies in the literature have been large psychiatric hospital surveys using buccal smears to determine X chromosome numbers. Some of these studies have suggested that Klinefelter's syndrome (XXY) is more prevalent among psychiatric patients than in the general male population, and similar results have been obtained for trisomy X (see tables 1 and 2). Regardless of this increased prevalence in some reports, the data show that X chromosomal abnormalities still only account for a very small proportion of individuals with schizophrenia or other psychoses.

Table 1. XXY in male psychiatric hospital patients.
XXY is present in .09% of live male births [8].

Reference:	# Patients with XXY:	Diagnosis of XXY Patients:
Tedeschi & Freeman 1962 [9]	3/248	Schizophrenia
Raphael & Shaw, 1963 [10]	1/210	Schizophrenia
Nielsen & Fischer, 1965 [11]	3/14 hypogonadal	Not schizophrenia
Judd & Brandkamp, 1967 [12]	1/22	Schizophrenia
Anders et al.,1968 [13]	6/529	Chronic psychosis
MacLean et al.,1968 [14]	30/6000	13 of 30,
Dasgupta et al.,1973 [15]	2/500	Schizophrenia
Trixler et al., 1976 [16]	2/310	Not schizophrenia
Axelsson & Wahlstrom, 1984 [17]	2/134	Paranoid psychosis
DeLisi et al.,1986 [18]	0/34	Only schizophrenia

50/8459 Psychiatric patients
(.59%)

Table 2. XXX in female psychiatric hospital patients.
XXX is present in .07% of live female births [8].

Reference:	# of Patients with XXX:	Diagnosis of XXX Patients:
Raphael & Shaw, 1963 [10]	1/105	Schizophrenia
Asaka et al., 1967 [19]	2/424	Schizophrenia
Anders et al., 1968 [13]	1/445	Chronic psychosis
MacLean et al., 1968 [14]	17/6241	7 of 17, schizophrenia
Kaplan, 1970 [20]	12/1061	Schizophrenia
Vartanian & Gendelis, 1972 [21]	9/2431	Schizophrenia

42/10,707
(.39%) Female psychiatric patients

Of the studies reporting complete karyotype analyses of samples of unrelated patients, some abnormalities other than those attributed to the X chromosome have been noted (see table 3). Among these is the finding of high numbers of acentric chromosomal fragments, which are common events events that probably occur postnatally, either as a result of cellular damage from radiation, chemicals, viral infections, extreme changes in cellular environment, or possibly due to no apparent cause other than effects of the normal aging process [22]. One study [17] describes an unusually high percentage of aberrations among patients with paranoid psychosis; these include long Y chromosomes, duplications in the heterochromatin regions of several chromosomes, inversion of a region on chromosome 9, and fragile sites on chromosomes 17 and 9.

Table 3. Studies of complete chromosmal analyses in schizophrenic patients

Reference:	# of Patients with aberrations	Diagnosis of patients:
Book et al., 1963 [23]	0/10	Childhood schizophrenia
Raphael & Shaw, 1963 [10]	1/10 (XXY)	Schizophrenia
Judd & Brandcamp, 1967 [12]	1/40 (XY/XXY)	Schizophrenia
Anders et al., 1968 [13]	11/54 (Acentric chromosome fragments)	Chronic psychosis
Kaplan, 1970 [20]	18/50 (Acentric chromosome fragments)	Schizophrenia
Axelsson & Wahlstrom, 1984 [17]	44/ 134 (Misc.)	Paranoid psychosis
DeLisi et al., 1986 [18]	0/34	Schizophrenia

With regard to associations of schizophrenia with "fragile sites" on chromosomes (areas that appear partially dismembered from the main body of the chromosome under specific culture conditions), there are some interesting descriptions. Fragile X is now known to be responsible for X linked mental retardation in males (the Martin-Bell Syndrome) [24,25] and possibly a significant proportion of childhood autism in males [26]. Psychotic symptoms appear frequently among the clinical characteristics of fragile X individuals. There are also reports of schizophrenia and its spectrum disorders among family members of fragile X mentally retarded males [27]. One examination, however, of 34 consecutively seen unrelated males with schizophrenia, failed to detect any evidence of fragile X chromosomes [18].

A fragile site on the long arm of chromosome 3, and one on chromosome 19 have also been seen in schizophrenic members within isolated families [28,29]. Another case report of trisomy 8 and schizophrenia, as well as the most recent presentation of a family with two schizophrenic members with an unbalanced translocation of the tip of the long arm of chromosome 5 onto chromosome 1 [30,31], indicate abnormalities in regions of the genome that should suggest future foci for attention.

Although no X chromosome linkages have been reported for schizophrenia, two independent groups now find a linkage of bipolar affective disorder to the end of the long arm of the X chromosome using color-blindness as a marker [32] and RFLP studies [33]. A failure to find X linkage in affective disorder, however, was also reported by another group of investigators [34, 35].

PROTEIN POLYMORPHISMS:

Prior to the widespread application of the newer DNA technology, protein polymorphisms, particularly red blood cell group and HLA antigens on lymphocytes were focussed on because of their high degree of polymorphism. These were studied in several disease states. With respect to schizophrenia, the data is inconsistent and varies among different geographic groups studies. Across studies, there appears to be some modest trend for an association of the A-9 HLA antigen with paranoid schizophrenia [36], and at least one study showing an association with HLA-DR8 [37]. While statistically significant, these associations do not appear to indicate a major role for HLA locus defects as a risk factor for schizophrenia, and there is no hypothesis about schizophrenia that would be be consistent with these spurious associations. Linkage of the HLA locus (on chromosome 6) with schizophrenia, although initially suggested by one study [38], has been ruled out in two other more comprehensive examinations [39,40].

Advances in the sensitivity of 2-dimensional gel electrophoresis now allow the detection of many more protein polymorphisms. These are just now being developed for application to human studies of inherited protein patterns.

DNA POLYMORPHISMS:

With the development of recombinant DNA techniques for the study of the human genome, an almost unlimited source of polymorphisms in the form of RFLP's (restriction fragment length polymorphisms) has become available. In addition, the further knowledge that hypervariable regions exist in multiple areas of the genome in the form of variable number tandem repeats (VNTR's) has enabled some investigators to synthesize probes that will identify these regions by standard hybridization techniques [41]. These, along with probes for known "candidate" genes, will be useful in future studies.

The candidate gene approach (table 4), rather than systematically scaning the genome using known highly polymorphic regions, has some advantage in that it supplies the methodology to test previously proposed biochemical hypotheses for schizophrenia. Among these are the dopamine, noradrenergic, serotonin, neuropeptide, and neuroenzyme hypotheses (see refs 42 and 43 for review).

Table 4. The Candidate Gene Approach to genetic studies of schizophrenia. Potential candidates and their chromosome locations [44] are listed.

Candidate:	Chromosome Location:	Known RFLP's:
Neuroenzymes:		
Monoamine oxidase (MAO)	X	No
Catecholamine-O-methyl transferase	22	No
Dopamine B-hydroxylase	9	No
Tyrosine hydroxylase	11	Yes
Phenylalanine hydroxylase	12	Yes
Adenosine deaminase	20	Yes
Neuropeptides:		
CCK	3	No
VIP	6	No
NPY	7	Yes
Neurotensin	--	No
Somatostatin	3	Yes
Enkephalin	--	Yes
Substance P	7	Yes
Gastrin releasing peptide	18	Yes
Receptors:		
Dopamine D_1 and D_2	--	No
Serotonin	--	No

Alpha-adrenergic	--	No
Beta-adrenergic	5	No
Miscellaneous:		
Huntington's Disease probe	4	Yes
Long arm of chromosome 5	5	Yes
Nerve growth factor	17	Yes
Other growth factors/oncogenes	--	Yes
Fragile sites	X,3,19	Yes

METHOD TO DETECT RFLPS:

Human genomic DNA is isolated from whole blood or transformed lymphocyte cultures in a series of extraction
steps. The DNA is then cut into fragments by restriction
endonucleases, separated, based on size, by gel electrophoresis, then transferred (and immobilized) to nylon
or nitrocellulose filters using the method of Southern
[45]. Phosphorous-32 labelled probes isolated from vec-
tors are incubated with prepared blots of DNA from
several individuals. Restriction fragments with complementary sequences to portions of the probe are visual-
ized by autoradiography. These methods have now become
standardized and are reviewed in detail by Maniatis et
al. [46].

RFLP ASSOCIATION STUDIES:

The first studies of schizophrenia and manic-depressive
illness were reported by Feder et al. with regard to the
the pro-opiomelanocortin (POMC) gene [47]. Although
opiates were hypothesized to be related to major psychiatric illnes and likely candidates for gene defects,
no associations or linkages were found using described
RFLP polymorphisms in this gene, which is located on
the 4th chromosome.

Another series of association studies was done by Detera-
Wadleigh and colleagues [48, 49]. While some initial
associations in small numbers of patients were reported
with specific polymorphisms in neuropeptide Y [48] and
adenosine deaminase (see table 5) [49], the former was
ruled out in a larger study, while the latter has not
yet been confirmed in a larger population. No associations were found to any known polymorphisms in the
following other gene loci: somatostatin, VIP, substance
P, complement components, insulin, gastric releasing
peptide, phenylalanine hydroxylase, and HRAS-1.

Table 5. Association data for the adenosine deaminase locus (adapted from Detera-Wadleigh et al. ref. 49; Chi-square = 4.09; p<.05).

Restiction Enzyme	RFLP Alleles (Kb's)	Frequency of Polymorphism	
		Normals	Schizophrenics
PvuII	4.8,3.8/ 4.8,3.8	23 of 24 (95.8%)	22 of 29 (75.9%)
	(4.8,3.8/ 3.3)	1 of 24 (4.2%)	7 of 29 (24.1%)
	(3.3/3.3)	0	0

RFLP LINKAGE STUDIES:

A linkage of manic-depressive disorder to a region on chromosome 11 close to the insulin, H-RAS-1, and tyrosine hydroxylase genes has been reported in the Pennsylavania Amish population [50]. The results from these pedigrees can be considered relevant to schizophrenia, since schizoaffect-ive disorder appears within the pedigrees found to have the chromosome 11 linkage, and it is of some note that many of the patients with bipolar disorder, at some time in their past were diagnosed as schizophrenic. In two studies by other investigators, chromosome 11 linkage was ruled out in a large Icelandic pedigree [51], and three non-Amish Amer-ican pedigrees with multiple mixed affective disorder di-agnoses [52]. Further chromosome 11 studies in high density schizophrenia families are in progress (Detera-Wadleigh et al., unpublished data), although schizophrenia and schizo-affective disorder were present in some of the family members of the bipolar pedigrees already investigated.

The phenylalanine hydroxylase (PAH) locus has been of particular interest since schizophrenic patients have been reported to have abnormalities in phenylalanine metabolism that may indicate decreased activity of PAH [53]. In addition, although not systematically studied, psychoses have been reported to be more frequent among family members of patients with PKU [54,55], leading some investigators to speculate that schizophrenics could be heterozygotes for PKU [53]. Using the polymorphisms in the locus for PAH already identified by Woo and colleagues [56,57], linkage of schizophrenia to PAH was indeterminable in 3 families studied thus far due to a lack of highly polymorphic alleles within these families (Detera-Wadleigh et al., unpublished data). These studies, however, are continuing.

CONCLUSION:

While a genetic predisposition for schizophrenia and affective disorder has long been known, it has only been recent that progress has been made in the search for specific gene markers. Manic-Depressive illness has been linked to an area of chromosome 11 in one population, and the X chromosome in another, leading investigators to propose that there may be several forms of psychotic illness. Schizophrenia may be genetically part of the spectrum of these disorders; however, no specific linkages have been reported in families where schizophrenia is predominant. There are some families where chromosomal aberrations appear in association with schizophrenia, and these might provide clues for specific locations of the "schizophrenia gene", while not implying that a substantial amount of schizophrenia is due to these aberrations (i.e. fragile X, translocated chromosome 5, etc.).

The slow production of research findings on the genetics of schizophrenia may be due to the difficulties involved in executing these investigations, given the nature of the disease process itself. Despite the reports that first degree relatives of schizophrenic patients have an approximate 10% chance of developing the illness in their lifetime [2], there are very few families to be found with more than one ill member. The difficult ascertainment of these families for genetic marker studies could result from the social aspects of the illness. Families of schizophrenics tend to be fragmented; ill members are not spoken of and are often distant from family relations. The ill members are often suspicious of research efforts, refuse participation, or unavailable for study. They often do not have offspring available, since the illness onset is early in the reproductive years. The question of how this illness survives in the face of reduced reproduction may be answered by future genetic studies delineating its relationship to affective disorder and the nature of the gene itself [3,58].

The application of molecular genetic techniques to studies of schizophrenia has just begun. Future systematic analysis of the sequenced genome may eventually lead to one or more loci that determine susceptibility to this illness.

REFERENCES

1. Kraepelin E (1907). <u>Lehrbuich der Psychiatrie</u>, Translated by Diefendorf AR. (New York, The MacMillan Company).
2. Gottesman II and Shields J (1982). <u>Schizophrenia: The Epigenetic Puzzle</u>. (Cambridge, Cambridge University Press).
3. Crow TJ (1986). The continuum of psychosis and its implication for the structure of the gene. British J. of Psychiatry, 149, 419.
4. Tsuang MT, Winokur G, Crowe RR (1980). Morbidity risks of schizophrenia and affective disorders among first-degree relatives of patients with schizophrenia, mania, depression, and surgical conditions. British J. of Psychiatry, 137, 497.
5. Gershon ES, DeLisi LE, Maxwell ME, et al.(in press). A family study of schizophrenia and schizo-affective disorder. Archives of General Psychiatry.
6. Angst J, Felder W, Lohmeyer B (1979) Schizoaffective disorders: Results of a clinical investigation, I. J. of Affective Disorders, 1, 139.
7. Kendell RE and Brockington IF (1980). The identifi-cation of disease entities and the relationship between schizophrenia and affective psychoses. British J. Psychiatry, 137, 324.
8. Hamerton JL, Canning N, Ray M, Smith S (1975). A cytogenetic survey of 14,069 newborn infants I. Inci-dence of chromosome abnormalities. Clinical Genetics, 8, 223.
9. Tedeschi L and Freeman (1962). Sex chromosomes in male schizophrenics. Archives Gen. Psychiatry, 6, 17.
10. Raphael T and Shaw MW (1963). Chromosome studies in schizophrenia. JAMA, 183, 1022.
11. Nielson J and Fisher M (1965). Sex chromatin and sex chromasome abnormalities in male hypogonadal mental patients. Brit. J. of Psychiatry, 111, 641.
12. Judd LL and Brandkamp WW (1967). Chromosome anal-yses of adult schizophrenics. Arch. Gen. Psychiatry, 16, 316.
13. Anders JM, Jagiello G, Polani PE, et al. (1968). Chromosome findings in chronic psychotic patients. Brit. J. of Psychiatry, 114, 1167.
14. MacLean N, Court-Brown WM, Jacobs PA, et al. 1968). A survey of sex chromatin abnormalities in mental hospitals. J. Med. Genetics, 5, 165.
15. Dasgupta J, Dasgupta D, Balasubrahmanyan M (1973). XXY syndrome XY/XO mosaicism and acentric chromosomal fragments in male schizophrenics. Ind. J. of Medical Research, 61, 62.

16. Trixler M, Kosztolanyi G, Mehes K (1976). Sex chromosome aberration screening among male psychiatric patients. Arch. Psychiat. Nervenkr., 221, 273.
17. Axelsson R and Wahlstrom J (1984). Chromosome aberrations in patients with paranoid psychosis. Hereditas, 100, 29.
18. DeLisi LE, White B, Reiss A, Gershon ES (1986). Cytogenetic studies of schizophrenic patients. In Shagass et al. (eds.) Biological Psychiatry 1985. p. 64. (New York, Elsevier Science Publishing Co.).
19. Asaka A, Tsuboi T, Inouye E, et al. (1967). Schizophrenic psychosis in triplo-X females. Folia Psychiat. et Neurol., 21, 271.
20. Kaplan AR (1970). Chromosomal mosaicisms and occasional acentric chromosomal fragments in schizophrenic patients. Biological Psychiatry, 2, 89.
21. Vartanian ME and Gindilis VM (1972). The role of chromosomal aberrations in the clinical polymorphism of schizophrenia. International J. of Mental Health, 1,93.
22. Levitan M and Montagu A (1977). Textbook of Human Genetics, p. 107, (New York, Oxford University Press).
23. Book JA, Nichtern S, Gruenberg E (1963). Cytogenetical investigations in childhood schizophrenia. Acta Psych. Scand., 39, 309.
24. Martin JP, Bell J (1943). A pedigree of mental defect showing sex-linkage. J. of Neurology and Psychiatry, 6, 154.
25. Opitz JM and Sutherland GR (1984). Conference report: International workshop on the fragile-X and X-linked mental retardation. American J. of Medical Genetics, 17, 5.
26. Blomquist HK, Bohman M, Edvinsson SO, et al.(1985). Frequency of the fragile-X syndrome in infantile autism. Clinical Genetics, 27, 113.
27. Reiss AL, Hagerman RJ, Vinogradov S, et al. (in press). Psychiatric disability in female carriers of the fragile X chromosome. Arch. Gen. Psychiatry.
28. Rudduck C and Franzen G (1983) A new heritable fragile site on human chromosome 3. Hereditas, 98, 297.
29. Chodirker BN, Chudley AE, Ray M et al. (1987). Fragile 19p13 in a family with mental illness. Clinical Genetics, 31, 1.
30. Sperber MA (1975). Schizophrenia and organic brain syndrome with trisomy 8 (group-C trisomy 8[47,XX,8+]). Biol. Psychiatry, 10, 27.
31. Bassett AS, McGillivray RC, Pantzar JT (1987). Autosomal abnormality linked to schizophrenia. New Research Abstracts. American Psychiatric Association, 140th Annual Meeting, Chicago, May 9–14, 1987, p. 61, NR75.
32. Baron M, Rish N, Hamberger R et al. (1987) Genetic linkage between X-chromosome markers and bipolar affective illness. Nature, 326, 289.

33. Mendlewicz J, Simon SP, Sevy S, et al. (1987) Polymorphic DNA marker on X chromosome and manic depression. The Lancet, i, 1230.
34. Gershon ES, Mendlewicz J, Gaspar M et al. (1980). A collaborative study of genetic linkage of bipolar manic depressive illness and red/green color blindness. Acta Psychiatr. Scand., 61, 319.
35. Gershon ES, Targum SD, Matthysse S, Bunney WE Jr. Color blindness not closely linked to bipolar illness. Archives General Psychiatry, 36, 1423.
36. Goldin LR, DeLisi Le, Gershon ES (1987). Genetic aspects to the biology of schizophrenia. In: Henn FA and DeLisi LE (eds.), Neurochemistry and Neuropharmacology of Schizophrenia, pp. 467487 (Amsterdam, Elsevier).
37. Miyanaga K, Machiyama Y, Juji T (1984). Schizophrenic disorders and the HLA-DR antigens. Biological Psychiatry, 19, 121.
38. Turner WD (1979). Genetic markers for schizotaxia. Biological Psychiatry, 14, 177.
39. McGuffin P, Festenstein H, Murray R (1983). A family study of HLA antigens and other genetic markers in schizophrenia. Psychological Medicine, 13, 31.
40. Goldin LR, DeLisi LE, Gershon ES (1987). The relation of HLA to schizophrenia in 10 nuclear families. Psychiatry Research, 20, 69.
41. Jeffreys AJ, Wilson V, Thein SL (1985). Hypervariable 'minisatellite' regions in human DNA. Nature, 314, 67.
42. DeLisi LE and Wyatt RJ (1985). Neurochemical aspects of schizophrenia. In: Lajtha A (ed.). Handbook of Neurochemistry, Vol. 10, pp. 553-587. (New York, Plenum Publishing Co.).
43. Henn FA and DeLisi LE (eds.) (1987). Neurochemistry and Neuropharmacology of Schizophrenia. (Amsterdam, Elsevier).
44. Howard Hughes Medical Institute, Human Gene Mapping Library. (January 1987), Chromosome Plots: Regional Localization of DNA segments on Human Chromosomes, 2.
45. Southern E (1975). Detection of specific sequences among DNA fragments separated by gel electrophoresis. J. Mol. Biology, 98, 503.
46. Maniatis T, Fritsch EF, Sambrook J (1982). Molecular Cloning: A Laboratory Manual. (New York, Cold Spring Harbor Laboratory).
47. Feder J, Gurling HMD, Darby J, Cavalli-Sforza LL (1985). DNA restriction fragment analysis of the proopiomelanocortin gene in schizophrenia and bipolar disorders. American J. of Human Genetics, 37, 286.
48. Detera-Wadleigh S, DeLisi LE, Berrettini WH, et al. (1986). DNA polymorphism in schizophrenia and affective disorders. In:Shagass C et al.(eds.): Biological Psychiatry, 1985. Proceedings of the IVth World Congress of Biological Psychiatry. (New York, Elsevier).

49. Detera-Wadleigh, de Miguel C, Berrettini W, et al. (in press). Neuropeptide polymorphisms in affective disorder and schizophrenia. J. of Psychiatric Reasearch.
50. Egeland JA, Gerhard DS, Pauls et al. (1987). Bipolar affective disorders linked to DNA markers on chromosome 11. Nature, 325, 783.
51. Hodgkinson S, Sherrington R, Gurling H, et al. (1987). Molecular genetic evidence of heterogeneity in manic depression. Nature, 325, 805.
52. Detera-Wadleigh SD, Berrettini WH, Goldin LR (1987). Close linkage of C-Harvey-ras-1 and the insulin gene to affective disorder is ruled out in three North American pedigrees. Nature, 326, 289.
53. Wyatt RJ, Gillin JC, Stoff IM, et al. (1977). Beta-phenylethylamine and the neuropsychiatric disturbances. In: Usdin E, Barchas J, Hamberg D (eds.), Neuroregulators and Neuropsychiatric Disorders., pp. 31-45 (New York, Oxford University Press).
54. Lippman (1958). The significance of heterozygosity for hereditary metabolic errors related to mental deficiency (oligomentia). American J. Ment. Defic. 63, 320.
55. Monro TA (1947). Phenylketonuria: Data on 47 British families. Ann. Eugenic, 14,60.
56. Woo SLC, Lidsky AS, Guttler F Et al. (1983). Cloned human phenylalanine hydroxylase gene allows prenatal diagnosis and carrier detection of classical phenylketonuria. Nature, 306, 151.
57. Daiger SP, Lidsky AS, Chakraborty R et al. (1986). Polymorphic DNA haplotypes at the phenylalanine hydroxylase locus in prenatal diagnosis of phenylketonuria. Lancet i, 229.
58. Crow TJ (1987). Psychosis as a continuum and the virogene concept. British Medical Journal Bulletin, 43 (3), 754.

15
Amish study of genetics of affective disorders

J.A. Egeland

INTRODUCTION TO AMISH STUDY AND SETTING

This paper will review the epidemiologic and genetic studies of major affective disorders among the Old Order Amish of Pennsylvania. The Amish Study has been funded for a 14 year period (1976-1990) by the National Institute of Mental Health. It was designed and directed by J.A. Egeland, who began a series of a studies on the Amish in 1959 and has developed close ties to the community over time. The academic base of the study has been the Department of Psychiatry of the School of Medicine, University of Miami, Florida. The work has fostered an interface between psychiatry, epidemiology, human genetics and molecular biology. There has been collaboration with eleven schools of medicine and a number of laboratories providing the research teams needed to address different clinical and genetic studies.

The long term objective of the research has been to clarify the genetic status of bipolar I, or manic-depressive, disorder. The study setting is uniquely suited to this goal [1]. The Old Order Amish population, located in southeastern Pennsylvania, is a genetic isolate. This settlement has remained endogamous and essentially a closed gene pool since the original pioneers arrived in America in the 18th century. The opportunity to test genetic hypotheses and to identify homogeneous subtypes of affective disorder, a heterogeneous condition, was enhanced in this group.

The European origins of the Amish trace to the Anabaptist movement in the 16th century. Seen as a radical party of the Reformation, the Anabaptist movement sprang up simultaneously in Switzerland and southern Germany. The Amish descend from the Swiss Brethren with early leaders from Zurich, Bern and environs. The group dispersed throughout southern Europe as a result of widespread persecution. There

are, however, extensive genealogies that permit
tracing the family lines and inter-marriages to their
European origins.

Besides these genetic origins, there are other
features of the population that make it ideal for
medical-genetic study. The Amish have unusually large
families extending to a fourth generation. There has
been a paucity of large families for the study of
manic-depressive disorder in the United States, making
the Amish a treasured resource. The few family members
not staying within the church tend to reside in the
same area and are available for study. Questions of
establishing paternity do not arise in this group to
confound results of genotyping.

The people maintain an agrarian life-style where
extended families live and work together. The
cohesiveness of church membership also encourages close
interaction between people. This guarantees multiple
informants and a richness in observations regarding any
psychiatric problem. It is true for historic, as well
as current, cases of illness within families. Finally,
the diagnosis of manic-depressive illness is not
complicated by the presence of alcoholism, drug abuse
or violence. Powerful cultural and religious
prohibitions of these behaviors mean they are less
likely to mask either the ascertainment or diagnosis of
unipolar and bipolar disorders.

PHASE 1:   DEMOGRAPHY AND EPIDEMIOLOGY (1976-1987)

The Amish population numbers over 12,500 and covers
more than a 500 square mile area. Families are divided
geographically according to church districts and a
"scribe", or informant, was designated for each
district. Canvassing all households through the scribe
network, a complete family census was made on
demographic characteristics of the population. The
Miami staff next conducted a community-wide
epidemiologic survey to ascertain active cases of
mental illness for the time frame of the study and to
record inactive cases of serious mental illness and all
suicides for a 100 year period. Up-dating of these
registries continues.

The first report of diagnoses for active cases of
psychiatric illness between 1976-80 (n=112) showed that
71% were either a bipolar (34%) or unipolar (37%)
affective disorder [1]. A more recent reference
included ascertainment through 1985 with 173 patients
diagnosed as actively ill since the study began [2].
Of that number, 62% were diagnosed as major affective
disorders, either a bipolar or unipolar form. There
were 32 cases of a bipolar I disorder, 19 instances of
bipolar II disorder and 56 cases of recurrent major

depressive disorder. The up-dated diagnostic breakdown over a ten-year period is shown in Table 1.

TABLE 1 Diagnoses of active cases among Amish:1976-1987

Diagnosis	Number	Percent
Bipolar I and II disorder	59	29
Unipolar depression	68	33
Minor depression	18	9
Hypomanic disorder	3	1.5
Cyclothymic disorder	3	1.5
Schizoaffective	12	6
Schizophrenia	5	2.5
Paranoid disorder	3	1.5
Psychosexual disorder	5	2.5
Generalized anxiety	5	2.5
Obsessive-compulsive	4	2
Anorexia	3	1.5
Personality disorders	15	7
Other	3	1.5
TOTAL	206	101%

Table 1 indicates that the active case sample through 1987 consists of 206 subjects. There are 40 cases of bipolar I, 19 of bipolar II and 68 of major depressive disorder. This accounts for 127 of the 206 cases or 62% of the entire sample. The lifetime prevalence for major affective disorder is 2% for the Amish population.

Compared to the original report of active cases through 1980, the latest figures given in Table 1 cover a wider spectrum of diagnoses. (Children under age 15 have not been systematically surveyed so those diagnoses are not included). For example, ascertainment now reveals cases of generalized anxiety disorder, obsessive-compulsive disorder, anorexia and a wider range of personality problems. Also noted between 1980 and 1987 are small increases in several diagnostic categories, although the prevalence of these disorders remains low. Examples are: schizophrenia = 5 cases (compared to 4 by 1980); paranoid disorder = 3 (2 by 1980); sexual disorders = 5 (1 by 1980); and personality disorders = 15 (7 by 1980).

The ascertainment of certain of these conditions was in part a function of the intensive evaluation of bipolar I families. Secondly, project efforts to provide referral for treatment and maintenance of strict confidentiality made it easier for people to report conditions, such as psychosexual, than was the case in the earlier phase of epidemiological study.

It had been noted before that there were no significant differences in the rates of bipolar and unipolar disorders for males and females in this population [1]. The same pattern can be seen with the present figures. Among the 127 BP/UP cases shown in Table 1, there are 63 males and 64 females. The slight increase for a bipolar form of the illness among men (34 male/25 female) and for unipolar depression among women (39 female/29 male) is not statistically significant. Age-sex adjusted rates give the same results as there is a remarkably equal sex distribution within the population as a whole and specifically for those at risk (ages 15-55).

Another subject of interest pertains to the ratio of bipolar and unipolar illness. The Amish Study reported previously that this ratio was essentially 1:1 when cases of bipolar II disorder were added to the bipolar I sample [3]. The latest figures (Table 1) show a total of 59 bipolar cases compared to 68 unipolar cases. This continues to suggest that hypomanic episodes for persons suffering a major depression are more common than usually reported. The Amish Study policy of following patients between hospital admissions is an advantage for detecting hypomania. Excluding the bipolar II cases, our study yields slightly less than a 2:1 (UP/BP) ratio in contrast to the 4:1 ratio more typically reported from other studies.

Finally, an analysis of suicides (n=26) for a 100 year period of 1880 to 1980 showed that except for one person with a minor depression, all suicided persons had suffered either a major depressive or manic-depressive disorder. Here, again, there were as many bipolars as unipolars and the suicides were clustered in only a few affective disorder families [4]. The last suicide occurred just prior to the start of the study and our field staff stays alert to an assessment of suicide risk.

Bipolar sample

The first 32 bipolar I subjects identified served to define the bipolar sample upon which the on-going genetic studies have been based. These 32 BP I probands marked entry into large families averaging 9.5 first degree (n=303) and 18 second degree (n=565) relatives. Considerable time was invested over several years to develop rapport and full cooperation from these extended family lines. All 32 BP patients and their relatives were interviewed using a structured psychiatric interview called the Schedule for Affective Disorders and Schizophrenia, Lifetime Version (SADS-L). Different staff members did multiple interviews including not only responses from the patient but also from a selected number of closest family members (spouse/parents/sibs/children) friends or clergy. Furthermore, all psychiatric hospital and clinic records for these bipolar subjects and their affected relatives were abstracted by a standard procedure.

Psychiatric evaluations were performed by a panel of five diagnosticians. They remained blind to proband status, all family relationships, previous diagnoses, treatment or treatment response. The clinicians noted no differences in sympmtomatology or course of illness for Amish subjects in this study compared to non-Amish patients in their general clinical experience. The interview materials and medical record abstractions were separated in time and intermixed with other cases as one type of reliability check using two sources of information. High levels of diagnostic reliability have been demonstrated [5]. There has been a periodic up-dating of active bipolar I cases.

PHASE 2: GENETIC LINKAGE STUDY BY CONVENTIONAL MARKERS (1976-1982)

Having completed diagnostic evaluation for the bipolar pedigrees, the Amish Study began an extensive search for a genetic marker using most of the conventional blood typings. Five bipolar families were chosen because of their structure and the segregation of affective disorders in a pattern that appeared informative for linkage. This collaboration involved Miami, Yale, Miller Memorial Blood Center, Pfizer Diagnostics, American Red Cross, Penn State University Medical Center and the UCLA School of Medicine.

A total of 94 family members were tested for color blindness and were typed for the Xg blood group. They were also typed for 42 red blood cell and related markers in 17 antigen systems. In the final phase of testing, fifty-nine individuals in two other pedigrees were further typed for the human leukocyte antigen (HLA) Bf and GLO. The results of the various genetic linkage analyses were combined in a single report [6].

Important to the original commitment of the Amish Study were those analyses assuming an X-linkage hypothesis. It had been an initial goal of the research to attempt replication of the widely publicized findings of linkage between affective disorder and the colorblindness and Xg blood group marker loci. The Amish pedigree data did not support X-linkage to either loci. The red cell Xg locus proved to be uninformative for the Amish, as in other studies. Magnitudes of the most negative lod scores for colorblindness were greater than those of the most positive, using a variety of diagnostic schemes.

X-linkage has not been supported in other investigations and the deutan/protan and Xg linkage hypotheses were shown to be incompatible because of the mapping distance between these loci [7]. Some studies continue to suggest X-linkage in selected families for the markers of colorblindness and 6GPD deficiency [8]. Further confirmation awaits work on the molecular level using DNA probes for this region on the distal long arm of the X chromosome.

As part of the conventional linkage study for Amish bipolar pedigrees, analyses were also done assuming autosomal dominant inheritance and testing 12 autosomal markers. The typed markers included: ABO, COMT, Rh, Duffy, Kidd, MNSs, P_1, Lewis, Yk^a, Cs^a, McC^a, and Kn^a. Again, a wide variety of diagnostic schemes was employed in the linkage analyses with LIPED. A few weak positive lod scores were noted (eg. Yk^a, Kn^a, ABO and MNS) but strong linkage of affective disorder was not highly supported for any of the marker loci.

Lastly, linkage analyses were conducted with HLA and GLO for two bipolar Amish pedigrees. The evidence was against close linkage with markers in this region of chromosome 6. Exclusion data from other studies with large families likewise rejected the notion of genetic linkage between HLA and major affective disorders [9].

PHASE 3: MODE AND PATHWAY OF INHERITANCE (1976-1986)

Segregation study

Concurrent with genetic linkage analyses, the sample of 32 bipolar pedigrees was examined by segregation analysis to test for specific genetic hypotheses on mode of transmission (Dr. Pauls at Yale). Tests were done in an hierarchial fashion beginning with "no genetic transmission" and then testing for "polygenic" and "single gene" models. The most likely model was shown to be that of an autosomal dominant mode of transmission with a moderately high penetrance of 63%. The diagnostic scheme which fit this model included bipolar I, schizoaffective disorder (manic or depressive types), bipolar II, atypical bipolar disorder and major depressive disorder. These particular diagnoses appeared to be related and to share a common genetic diathesis. The segregation analyses reinforced the search for genetic markers linked to major affective disorders [10].

Progenitor trace study

A separate question from that of the mode of genetic transmission has been one seeking proof of the historical gene pathways through the generations. Since the inception of the Amish Study, an effort has been underway to trace the possible point of introduction of the putative bipolar gene or genes into the group.

The Amish are excellent observers of family patterns and have always asserted that there has been a clustering of psychopathology within certain family lines. Dating from work among the Amish in the 1960's, it was further recognized that perhaps most patients with major affective disorders (by comparison to Amish persons in general) traced back to particular ancestors along particular pathways. Significant differences in progenitor profiles for the bipolar patients compared to controls might allow conclusions regarding the genetic homogeneity of confirmed cases of affective disorder among these descendants. The Progenitor Trace Study was designed to confirm or invalidate these impressions.

All Old Order Amish descend from the relatively few European "founders", numbering about 30, who immigrated to the original settlement in Pennsylvania during 1730-1770. The Progenitor Trace Study does indicate that only a few of the original pioneer couples are "likely candidates" for introduction of the bipolar gene. Evidence (family accounts, court, hospital and geneological records prior to 1900) reveals only a few families with serious psychiatric conditions (including instances of manic disorder) through successive generations beginning with children or grandchildren of the pioneer (unpublished paper). For most pioneer couples, no serious mental illness can be ascertained historically down through the first three to four generations. This allows the Amish Study to focus with increasing assurance on one of several progenitors who presumably introduced bipolar affective disorder into the settlement.

PHASE 4: GENETIC LINKAGE BY DNA METHODS (1982-1987)

When the Amish Study was initiated in 1976, no DNA polymorphisms in man had been detected. When the conventional blood typings were performed in 1979, polymorphism detection at the DNA level was still hampered by the lack of cloned probes. Then a great impetus to this effort occurred in the development and application of recombinant DNA methods. DNA polymorphisms would soon be generated at an unprecedented rate. Scientists predicted the mapping of the human genome within a decade. The time seemed right to apply DNA methods to the search for genetic markers linked to a predisposition to psychiatric disorders.

One large, multi-generational family (Pedigree 110) had been selected from the sample of 32 bipolar I pedigrees. Blood samples were drawn in 1982 from 51 members of Pedigree 110 and the research team embarked on full linkage study of all restriction fragment length polymorphisms (RFLP's) available for typing. The permanent lymphoblastoid cell lines from this pedigree were established at the Coriell Institute for Medical Research (IMR). By this means, Pedigree 110 (IMR Family 884) became the first large pedigree for bipolar affective disorder available to scientists for molecular genetic study [11].

Many collaborators contributed to the RFLP typings. The RFLP typings done at Yale (in Dr. K. Kidd's laboratory) made it possible to rule out many markers spanning the human genome [12]. One of the highest densities of RFLP's was on the short arm of chromosome 11 where extensive mapping work was taking place (in Dr. D. Housman's laboratory). Nine DNA loci were tested for Pedigree 110 and haplotypes assigned, blind to diagnosis, by Dr. D. Gerhard at the Massachusetts Institute of Technology. Linkage analyses suggested a major locus for affective disorders might exist on chromosome 11 [13]. The results were provocative but not definitive, with a LOD score of 1.7. Therefore, an additional 30 blood samples were collected in 1984 from other members of Pedigree 110. The sample then consisted of 81 persons, 19 with a major affective disorder and 62 unaffected.

Results published in NATURE gave evidence that DNA markers on chromosome 11 were strongly linked to a predisposition for major affective disorders (14). The genetic markers on chromosome 11 were insulin (INS) and the cellular oncogene Ha-ras-1 (HRAS1). The LOD scores had increased to 3.3 for HRAS1 and 1.7 for INS using the maximum penetrance value of .63 (ages 30 and over) as derived from segregation analysis of the entire bipolar sample. A LOD score of 3 or greater gave odds of at least 1,000 to one against this pattern having occurred by chance alone.

The maximum penetrance value was 0.85 for Pedigree 110 when analyzed separately from the entire sample of 32 bipolar families [10]. Using this penetrance value, the LOD scores for HRAS1 and INS increased to 4.1 and 2.6 respectively. When both marker loci (HRAS1 and INS) were considered simultaneously using multipoint linkage analysis, the LOD score was 4.9 (giving odds greater than 10,000 to one). The affective disorder gene was seen to be most closely linked to the HRAS1 locus on the short arm of chromosome 11. This was the first scientific report on the localization of a gene conferring a predisposition to a common psychiatric condition by means of recombinant DNA techniques.

PHASE 5: FUTURE GENETIC RESEARCH

The Amish Study team has embarked on the next steps that must be taken to carry genetic findings forward. There are five efforts that should be cited. The first pertains to a replication of the chromosome 11 findings among the Amish. To do this, blood samples have been obtained from another 30 members of Pedigree 110 and the haplotyping is underway to see if genetic markers in the HRAS1 region of 11p continue to show linkage to the illness. Cell lines also have been established from a second Amish bipolar pedigree (IMR Family 1075). This is a large family that should prove to be genetically homogeneous and hopefully show segregation for the same susceptbility gene. These Amish pedigree extensions are likely to provide additional informative meiotic events to help with the determination of the affective disorders locus [11].

Since replication of the chromosome 11 findings is needed for non-Amish bipolar families, one such family was ascertained and diagnosed in collaboration with Dr. D. Papolos (Albert Einstein College of Medicine). Permanent cell lines have been established for this family to detect whether there is linkage between the illness and the INS-HRAS1 chromosomal region for persons with distinct ancestral backgrounds from the Amish.

The second related effort is based on collaboration with Dr. D. Gerhard (now at Washington University School of Medicine) who is conducting RFLP study of the cell lines described previously. One direction of her work is the prediction of candidate genes, including interest in tyrosine hydroxylase (TH), an enyzme important in dopamine metabolism. The TH gene is mapped to the same region of chromosome 11 as INS-HRAS1 and may or may not be the putative gene. Another line of investigation requires further development of a physical map for the INS-HRAS1 region of chromosome 11 and isolation of additional markers in this region [15].

A third step planned for the Amish Study is to examine whether or not other disorders in the affective spectrum are linked to the same genetic markers. Data from Pedigree 110 indicated that four clinically variant forms of affective illness shared the common chromosome 11 haplotypes; namely: bipolar I, bipolar II, schizoaffective and major depressive disorders. It is not known if other and/or milder forms of psychiatric illness share the same genetic diathesis.

The fourth goal is to design an "at risk" prospective study of children in bipolar I families. Remaining blind to genotyping, one could follow "at risk" and control subjects over time to assess all factors, environmental as well as genetic, that might be associated with penetrance of the gene and the onset of the illness. An ability to identify individuals at high risk would be a step toward genetic counseling and early intervention.

Fifth, the Progenitor Trace Study will continue to employ historical traces, in addition to utilizing computer programs on the total Amish genealogy, in order to ascertain the gene pathways back in time. It may be possible to identify more than one subtype of bipolar affective disorder segregating in the Amish population. Furthermore, it may be possible to identify affective families in the United States or Europe whose biological ties with Amish bipolar progenitors might make them valuable candidates for molecular genetic study. The most exciting aspect of research into the genetics of affective disorders is growing support for this avenue of investigation involving multi-disciplines and new methodologies [16].

1.   Egeland, JA and Hostetter, AM (1983). Amish Study: I. Affective disorders among the Amish, 1976-1980. Am. J. Psychiatry 140:56-61.
2.   Egeland, JA (1986). Cultural factors and social stigma for manic-depression: The Amish Study. Am. J. Soc. Psychiatry 6:279-286.
3.   Egeland, JA (1983). Bipolarity: The iceberg of affective disorders? Comp. Psychiatry 24:4,377-344.
4.   Egeland, JA and Sussex, JN (1985) Suicide and family loading for affective disorders. J. Am. Med. Assoc. 254:915-918.
5.   Hostetter, AM, Egeland, JA and Endicott, J (1983). Amish Study: II. Concensus diagnoses and reliability results. Am. J. Psychiatry 140:62-66.
6.   Kidd, KK, Egeland, JA, Molthan, L, Pauls, DL, Kruger, SD and Messner, KH (1984). Am. J. Psychiatry 141:1042-1048.
7.   Gershon, ES, Targum, SD, Matthysse, S and Bunney, WE (1979). Colorblindness not closed linked to bipolar illness. Arch. Gen. Psychiatry 36:1423-1430.
8.   Baron, M, Risch, N, Hamburger, R, Mandel, B, Kushner, S, Newman, M, Drumer, D and Belmaker, RH (1987). Genetic linkage between X-chromosome markers and bipolar affective illness. Nature 326:289-292.
9.   Goldin, LR, Clerget-Darpoux, F and Gershon, ES (1982). Relationship of HLA to major affective disorder not supported. J. Psychiatry Res. 7:29-45.

10.   Pauls, DL and Egeland, JA (1988). Autosomal dominant inheritance of bipolar affective disorders: Segregation analyses of Old Order Amish families. (Submitted for publication)

11.   Egeland, JA.   Amish major affective disorders pedigrees, in <u>1985 and 1986/1987 Catalog of Cell Lines</u>, NIGMS Human Genetic Mutant Cell Repository. (eds.) NIH Publications 85-2011 and 87-2011.   Washington, DC. US Department of Health and Human Services, 1985, pp 560-567, 1986/1987, pp. 298-306 and 489-496.

12.   Kidd, JR, Egeland, JA, Pakstis, AJ, Castiglione, CM, Pletcher, BA, Morton, LA and Kidd, KK (1987). Searching for a major genetic locus for affective disorder in the Old Order Amish. J. Psychiat. Res. 21:577-580.

13.   Gerhard, DS, Egeland, JA, Pauls, DL, Kidd, JR, Kramer, PL, Housman, DE and Kidd, KK (1984). Is a gene for affective disorder located on the short arm of chromosome 11? Am. J. Hum. Genet. 36:3S.

14.   Egeland, JA, Gerhard, DS, Pauls, DL, Sussex, JN, Kidd, KK, Allen, CR, Hostetter, AM and Housman, DE. (1987). Bipolar affective disorders linked to DNA markers on chromosome 11. Nature, 325:783-787.

15.   Gerhard, DS, Egeland, JA, Pauls, DL and Housman, DE (1987) Search for a gene that predisposes individuals to BPI disorder. J. Psychiat. Res. 21:569-575.

16.   Gershon, ES, Merril, CR, Goldin LR, DeLisi, LE, Berrettini, WH and Nurnberger, JI (1987). The role of molecular genetics in psychiatry. Biol. Psychiatry 22:1388-1405.

16
Affective disorders segregation structures according to clinical and pharmacological features

E. Smeraldi, M. Gasperini, F. Macciardi, A. Orsini, M. Provenza, G. Sciuto and C. Bussoleni

Affective Disorders are etiologically heterogeneous when starting from clinical data and using different partioning criteria; we are not still able to discriminate among hypothetically and etiologically different forms of disease. We should identify unequivocal phenotypes strictly related to well shaped genetic structures or, using an opposite approach, we should validate phenotypic criteria with formal genetic methodologies (1). Among phenotypic criteria the subdivision into Endogenous and non Endogenous Depression failed to be validated from the point of family study data. In fact they did not confirm the hypothesis that first degree relatives of patients with Endogenous depression had a higher rate of depressive illness.

A suggestion could be derived as a working hypothesis when searching for the formal basis of genetics in Affective Disorders: we should look at data without any sound a priori assumption. We could begin to analyse how many relatives are affected, and which families they belong to, studying the segregation pattern of the disease in accordance with biologically clear cut criteria. In this aim, in previous studies (2,3) we developed genetic strategies to generate subgroups of families as homogeneous as possible in order to detect possible differential transmission modes of Affective Disorders. We faced the problem of how validly analysing the segregation structures of different families without determining identification and classification parameters for families being considered before, since we could not be sure that information derived from particular cases could correspond to unequivocal biological structures. Using pure statistical analyses (4,2) the outcome on Lithium therapy was previously confirmed to be an useful criterion for defining differences in susceptibility to Affective Disorders: in fact, whatever the polarity , the frequency of the disease was greater in first degree relatives of non relapsed probands (with good Lithium outcome) than in those of relapsed ones (with poor Lithium outcome). This could suggest that the two groups of families, according to proband's Lithium outcome, might be samples of two not completely overlapping populations.

In following studies we tested the pedigree of each family under two different Single Major Locus (SML) transmission hypotheses: a dominant one with a sex-effect (related to Lithium non relapsed probands) and a recessive one (related to Lithium relapsed probands) and we calculated the log-likelihood ratios (3). Since we did not test the two models but we only studied the fitting of data to the two alternative hypotheses, this ratio value did not indicate whether or not the mode of transmission involved were a SML one but only whether or not the distribution of secondary cases in the family were more suitable with the first or second model.

The outcome of probands on Lithium treatment appeared to be an useful discriminatory criterion for the identification of homogeneous subgroups of Affective Disorder, but it only provided a trend in this respect, so, in a following study (5), we applied the same strategy to a greater number of informative families independently of this variable and we obtained a distribution of log-likelihood ratios significantly different from the normal distribution which would be obtained only if a single genetic mechanism underlied the susceptibility to the disease. In fact we recognized at least three higher frequency peaks which represented the families with ratio values consistent with a dominant transmission, the families which might have a susceptibility system different from recessive vs. dominant division and families with a susceptibility system fitting the recessive model of transmission. This distribution was obtained considering only secondary cases with Major Affective Disorders (Bipolar Disorders and Major Depressions, Recurrent) according to DSM-III criteria (6). Since these were several experimental data (7,8,9) showing that diagnostic classes including Dysthymic Disorder, Cyclothymic Disorder and Atypical Depression, when occurring in informative families, could be incomplete forms (intermediate phenotypes) of the same disease.

In a following step we tested what kind of log-likelihood ratio distribution we would obtain by including these other diagnoses as affected phenotypes assuming that they all belong to the same diagnostic spectrum of Major Affective Disorders (10). Comparing the log-likelihood ratio values estimated after excluding or including the affective spectrum disorders we observed that their inclusion did not cause a single shift towards only one of the two genetic hypotheses of transmission. This finding could be due to the type of relationship of the affected relative in the pedigree.

An important finding was the recognition of significant effects of spectrum diagnoses on the pattern of the segregation analyses. Using spectrum disorders to detect secondary cases of depressive disease greatly modified the familial segregation patterns of our affective probands, and indicated the great importance of using intermediate phenotypes to have more precise indications of susceptibility to use in segregation studies (11).

Since for genetic studies it is of great interest to study the segregation of susceptibility to the disease and not the segregation of the disease per se, we used segregation analyses (12) to identify the familial structures of susceptibility whatever the phenotypical manifestations within the spectrum of Affective Disorders. In this sense, pure clinical diagnoses and external validity criteria would be provided and the study of the susceptibility mode of transmission allowed us to group disorders with symptomatological similarities and with different epidemiological characteristics, for instance recurrence and duration.

Among the available information about our probands and their relatives, two kinds of data might be useful to indicate the reality of the genetic heterogeneity observed: these are the outcome after therapy with antidepressant tricyclics (TCA) and the presence of Personality Disorders (PD).

It is well known that, despite the success of antidepressant tricyclics poor treatment outcome does occur in same patients. At this time, neither simple pharmacokinetic mechanisms nor clinical or semeiological patient characteristics, have been able to explain these negative outcomes. Since the presence or absence of a drug response may differentiate biologically and/or biochemically distinct forms of Affective Disorders that may be genetically (and thus etiologically) different, we started to test the relationship between the familial susceptibility segregation and the possible genetic mechanism underlying the proband outcome on antidepressant treatments (13). The results of that analysis are reported in Table 1. For the group of "tricyclic drug responders" the mendelian model appeared to be the most suitable. A good fit also was obtained under the SML additive transmission hypothesis. The model which best fitted the group of "poor responders" was the polygenic one, while for the "non responders" the best fit appeared to be the SML hypothesis with transmission probability estimate (for heterozygotic genotype). These findings confirmed that the pharmacological criterion of antidepressant tricyclic drug treatment outcome really allowed us to identify different familial segregation structures of Affective Disorders. Even the presence of Personality Disorders have been hypothesized as a possible criterion of heterogeneity of affective illness. Previous investigations had pointed out how personality traits and disorders were of great interest both in affecting the clinical picture and course of affective illness, and influencing pharmacological treatment outcomes. It is important to underline how the presence of specific Personality Disorders (14) could affect the outcome on antidepressant treatment in terms of a reduced improvement of symptomatological pictures. In other words, the cohexistence of an Axis II diagnosis and of an Affective Disorder on Axis I induces lesser degree of recovery from Major Depression.

Furthermore, certain personality disorders could be considered as predisposing factors for depressive illness or perhaps could be viewed as subclinical manifestations of milder forms of Affective Disorders.

Table 1. Maximum likelihood values estimated for each group of families according to different hypotheses of disorder transmission

Hypotheses	Tricyclics		
	Responders	Poor responders	Nonresponders
No transmission	-1016.47	-397.61	-101.31
SML recessive	-397.83	-157.35	-18.41
SML additive	-273.19	-114.11	-18.60
SML dominant	-317.76	-160.14	-15.45
Mendelian	-272.62*	-114.11	-18.41
TAU2			-1.73*
Polygenic	-281.52	-102.28*	-3.52

* Best fit solution
Population prevalences: male = 0.0091; female = 0.0102 (13)

A preliminary study on relationship between Affective and Personality Disorders in the Italian population (15) reported that the presence or absence of whatever Personality Disorder in probands predicted differential affective risks in families. Personality Disorders more represented were the Hystrionic (30%) in bipolar, Compulsive (32%), Avoidant (18%), Dependent (25%), Passive-Aggressive and Schizoid (11%) in unipolar patients while borderline personality disorder was more frequent in controls.

When statistically analysing the frequency of the illness in first degree relatives of unipolar and bipolar probands and of controls with any psychiatric disease with and without Personality Disorders a significant effect of the presence of Personality Disorders in the unipolar and control groups in enhancing the familial morbidity risks was found (Table 2). In other words, in the group of controls and in the one of unipolar patients the presence of Personality Disorders in the probands affected higher risks for depression in relatives.

All the reports could be consistent with the hypothesis that Personality Disorders might add to the severity of the form from the point of familial load and early development of the affective disorders.

Table 2. Logistic analysis: statistical significance of variables in predicting Affective Disorders frequencies in first-degree relatives of bipolars, unipolars and controls

	Z values
Baseline frequency of disease	−9.3899*
Presence or absence of PD in probands	−1.5526
Type of relationship	0.159
Sex of probands	1.4359
Comparison among bipolars, unipolars, controls	1.9236
	1.3637
Interaction bipolars, unipolars, controls/PD	3.5737*
	−0.715

* p=0.01; for three-level varaiable comparison among bipolars, unipolars and controls analysis estimated two Z values (15)

For this reason we realized to perform segregation analyses taking into account, in addition to the pharmacological criterion the presence or the absence of Personality Disorders to possibly singling out more hypothetically homogeneous study groups of affective patients (16). The patients previously defined as "poor responders" were furtherly partioned into "non responders" or "respoders" respectively according to their slight improvement while on antidepressant tricyclics therapy and/or their bad outcome on treatment with one tricyclic and a good outcome on tratment with another one. Since in the previous study (13) the group of families of "poor responders" probands, for whom the polygenic model was the best fitting, could be considered still heterogeneous.

The estimated likelihood values obtained (Table 3), allowed us to rule out the hypothesis of no genetic transmission for all the groups. For the group of "tricyclics responders with PD" the SML dominant model appeared to be the most suitable. The models which best fitted the group of "Non-responders with PD" were the dominant and the additive SML and the Mendelian without any valuable difference among them : moreover we could not rule out the Polygenic model. So the exact mode of transmission in these families seemed to be still unclear and the amount of heterogeneity was even high.

The dychotomic subdivision presence/absence of Personality Disorders seemed directly related to different genetic structures, the patients with PD fitting a saturated model of transmission in contrast to subject whose segregation pattern appeared best described by models characterized by a penetrance defect: we should remember that, up to now, we cannot say what penetrance defect might be effectively. Moreover this phenotypic characterization appeared more important than the outcome on tricyclics in eliciting homogeneous subgroups of disease.

Table 3. Goodnesses of fit for different groups of families according to alternative hypotheses of Affective Disorders transmission

	With Personality Disorders		Without Personality Disorders	
	Tricyclic Responders (n=26)	Tricyclic Non Resp. (n=17)	Tricyclic Responders (n=29)	Tricyclic Non Resp. (n=3)
HYPOTHESES:				
No transmission	1092.47	462.07	1343.4	61.03
Polygenic	648.11	223.25	580.07	31.52
Mixed	617.57		517.14	
Recessive SML	686.01	278.27	603.99	33.98
Dominant SML	466.37*	221.35*	486.03	17.74
Additive SML	467.40	221.35*	656.17	17.73
Mendelian SML	467.37	221.35*	486.64	34.37
Tau 1	467.39		173.29*	
Tau 2	488.71	232.13	496.19	17.42*

* Population prevalences: male=0.0091; female=0.0102 (16)

Personality Disorders appeared to be as useful as antidepressant tricyclics outcome in detecting the heterogeneity of Affective Disorders from the point of the familial structure. In fact, beside to the known reducing effect of clinical responses to therapy, the presence of Personality Disorders has been found to affect the frequency of Affective Disease in families, thus conditioning a differential mode of susceptibility transmission.

the specific ones, we did fail to identify a possible cluster of Personality Disorders truly linked to the susceptibility structure of Affective Disorders.

On dealing of such a phenotypical variability we obtained a better estimate of the underlying mode of transmission singling out different SML hypotheses, which, even though not yet conclusive, allowed us to rule out the Polygenic model. The choice of SML models appears to be of great interest when considering its suitability for linkage studies.

The complex genetic pattern that we just described must be analysed in its structural compounds to see whether or not the identified phenotypes are conceivably related to genetic determinants. We can hypothesize different frequencies of heterozygous subjects to be affected, leading to theoretical multiple threshold models; an indirect confirmation can be derived from diagnoses of spectrum and Atypical Affective Disorders which are not homogeneously distributed, since we found an excess of Dysthymic Disorders, Cyclothymic Disorders and Atypical Depressions among the relatives of the "tryciclic responders" independently from the existence of a particular personality trait in the probands. These so called intermediate phenotypes could also be incomplete or milder expression of the underlying genetic structure suggesting the extreme importance of a careful evaluation of the whole phenotype in its costituents.

The particular phenotypic characterization of subjects with PD vs subjects without PD appears to be more important than the outcome on tricyclics therapy from a genetic point of view: the responder/non responder dychotomy alone is completely informative to set up a particular segregation structure when conditioned by a Personality Disorder. This finding could have its own specific meaning, taking into account what several Authors have pointed out focusing on the complex relationship between the two variables. The presence of a Personality Disorder, which we evaluated when a given subject was normothymic, is not identifiable with the concept of premorbid personality, but it could even correspond to a not complete recovery under tricyclic therapy.

References

1) Andreasen N C, Scheftner W, Reich T, Hirschfeld R M A, Endicott J and Keller M B,(1986). The validation of the concept of Endogenous Depression. A Family Study Approach. Arch. Gen. Psychiatry, 43: 246–251

2) Smeraldi E, Petroccione A, Gasperini M, Macciardi F, Orsini A,

and Kidd K K (1984). Outcomes on lithium treatment as a tool for genetic studies in Affective Disorders. J. Affect. Dis., 6: 139-151

3) Smeraldi E, Petroccione A, Gasperini M, Macciardi F, and Orsini A (1984). The search for genetic homogeneity in Affective Disorders. J. Affect. Dis., 7: 99-107

4) Morabito A, Gasperini M, Macciardi F and Smeraldi E (1982). Possible relationship between outcome in primary affective disorders treated with lithium and family history. In: Costa E and Racagni G (eds.) "Typical and Atypical Antidepressants", Raven Press, New York, N.Y., pp. 157-163

5) Smeraldi E (1985). Genetic Aspects in Affective Disorders. Paper presented at EMRC Workshop "Prediction of course and outcome in Depressive Illness: Needed Areas for Research", Cork, Ireland, August 28-30

6) American Psychiatric Association (1980). Diagnostic and Statistical Manual of Mental Disorders, 3rd Edition, A.P.A., Washington D.C. (1980)

7) Hirschfeld R M A, Klerman G L, Andreasen N C, Clayton P J and Keller M B (1985). Situational major depressive disorders.Arch. Gen. Psychiatry, 42: 1109-1114

8) Perris C, Perris H, Ericsson U and Von Knorring L (1982). The genetics of depressions. A Family study of unipolar and neurotic reactive depressed patients. Arch. Psychiatr. Newenkz. 232: 137-155

9) Orsini A, Gasperini M, Smeraldi E et al. (, Genetic study of Affective Disorders in Italian and Swedish populations. In press.

10) Gasperini M, Orsini A, Bussoleni C, Macciardi F and Smeraldi E (1987). Genetic Approach to the Study of the Heterogeneity of Affective Disorders. J. Affect. Dis. 12,105-113 (1987).

11) Goldin L R, Gershon E S, Targum S D, Sparkes R S and Mc Ginnis M (1983). Segregation and linkage studies in families of patients with bipolar, unipolar and schizoaffective mood disorders. Am. J. Hum. Genet. 35: 274-287

12) Lalouel J M, Rao D C, Morton N E and Elston R C (1983). A unified model for complex segregation analysis. Am. J. Hum. Genet. 35: 816-826

13) Orsini A (1987). Antidepressant responses and segregation analyses in affective families. "Anxiety and Depression: Assessment and Treatment". Smeraldi E and Racagni G Eds,Raven Press, New York

14) Roose S P, Glassman A H, Walsh B T and Woodring S (1986). Tricyclic non responders, phenomenology and treatment. Am.J. Psychiatry 143: 373-374 (1986).

15) Gasperini M and Provenza M (1987). The relationship between Affective Illness and Personality Disorders. Preliminary Reports. "Anxiety and Depression: Assessment and Treatment". Smeraldi E and Racagni G Eds, Raven Press, New York

16) Smeraldi E, Gasperini M, Macciardi F, Orsini A, Provenza M and Sciuto G (1987) Segregation analysis in families of affective patients subdivided according treatment outcome and personality disorders. Paper presented at the Congress "New directions in Affective Disorders". Jerusalem, Israel, April 5-9

17
Psychiatric disorders in families of phenylketonuric patients

E. Smeraldi, F. Macciardi, M. Leporatti, O. Gambini, A. Orsini, C. Fara, R. Valsasina and E. Riva

INTRODUCTION

Phenylketonuria is an autosomal recessive disease caused by the inborn deficiency of phenylalanine hydroxylase enzyme in the homozygous individual. There are different defects in the enzymatic system of phenylalanine metabolism (1) and each of them can cause hyperphenylalaninemia, but we considered only the classical form of phenylketonuria. Usually the disease is associated with severe mental deficiency but there are several cases reported with near normal intelligence and some of them appeared to suffer from behavioral disturbances (2). Our aim was to select a sample of phenylketonuric patients and to study their familial pattern, particularly in their parents, in order to see whether there are any psychiatric pathologies in the supposed heterozygous carriers.

The rationale of this approach is that, according to a recent study of our Institute (3), it can be suggested that a genetic component of susceptibility for psychotic disorders might be heterozygosity for some alleles responsible for amino acid disorders in the homozygous state; in this sense heterozygosity for phenylketonuria could somehow bring about, even if in a lighter way than for the homozygotes, a clinic situation not only of intellective deficiency, but also of a psychotic or a psycho-affective pathology.

At present few information about the specific features of heterozygotes for phenylketonuria are available, both in psychopathology (4) and biology. Among several biochemical defects only some are completely known, such as for example the reduced tolerance for phenylalanine load (5) or the around 7-10% reduced activity of liver phenylalanine hydroxylase enzyme (6).

Such metabolic defects due to the heterozygous state suggest the possibility of an increased incidence of psychiatric disorders in families of patients with phenylketonuria. Moreover, when dealing

with the more typical abnormality due to PKU, that is the
retardation of mental development, we could hypothyze that, even
though heterozygotes do not show a quantitatively detectable defect,
the retardation could increase later during the life. In fact,
several structures involved in the pathogenesis of the disease (such
as myelin and dendritic spines (7,8)) continuously undergo to
dynamic processes and are quite susceptible to biological
unfavorable conditions.

MATERIALS AND METHODS

Heterozygous subjects have been ascertained by identification of
their children with phenylketonuria. Screening for phenylketonuria
has been carried out using Guthrie test. Phenylalanine plasma levels
higher than 20mg/100ml and abnormal urinary excretion of the
associated metabolites were present for all the patients. We
excluded from the study children with simple elevation of serum
phenylalanine. Heterozygous state for parents has been checked also
evaluating plasma phenylalanine/tyrosine molar rate (9).
So we selected a group of 80 parents, 38 male with mean current age
of 35 and 42 female with mean age of 31. Information about 1750
relatives has been collected using Family History, a semi-structured
interview that allow us to collect data about the familial pattern
of probands. It was possible to complete information about some
affected relatives consulting clinical records or family doctors.
Diagnosis of psychiatric disorders were obtained according to DSM
III (10). Information was collected for first, second, third and
when possible fourth degree relatives and psychiatric diagnosis of
interest were: schizophrenic disorders, alcoholism, affective
disorders, mental retardation, suicides and suicide attempts. A
group of 47 parents has been undergone to the WAIS test to determine
the verbal, performance and total intellective quotient. Before
interview and test, we had a conversation oriented to problems
generated by the affected child in the family to motivate the
participation of the parents in the interview and to decrease
anxiety and reticence before performing WAIS test.

RESULTS

The group of selected subjects, that is individuals who belong to
families in whom the dismetabolism clinically manifest as
psychiatric disorders, does not appear to be a random sample
representative of a specific population at risk for such diseases;
but it is a suitable sample to investigate the occurrence of

psychiatric disorders and the goal is the general understanding of the problem.

Table 1 summarizes our results: 50 of 1750 relatives that is 2.9% were affected by different psychiatric disorders.

Table 1. Affected relatives in 44 families of PKU patients

	n	%	M	F
Schizophrenia	11	0.6	9	2
Alcoholism	21	1.2	17	4
Affective Disorders	7	0.4	3	4
Mental Retardation	5	0.3	2	3
Suicide Attempts	3	0.2	1	2
Suicides	3	0.2	2	1
Total	50	2.9	34	16

Some considerations are of interest: parents we have located through their phenylketonuric children were mainly young people with a mean age of 33 and this does mean that, since the wide range of age at onset for psychiatric disorders, the probability to develop diseases is still high for the sample; moreover data we have taken into account were pooled and since it was not possible to estimate a precise age correction for each specific disorder, they were not corrected for age. For this reasons we realize that all prevalences reported are underestimated. In detail we found these frequencies for psychiatric disorders in our sample:
– for schizophrenia we have a risk of 0.6% which overlaps the lowest values expected in general population;
– for depression and other mood disorders we found a risk of 0.4% These low values compared with the one of general population could be due to the fact we did not better specified diagnosis of alcoholism. In fact this disorder is often correlated with affective spectrum pathology so that many cases could be included in this category. On the other hand the diagnosis of alcoholism is of not easy interpretation since the impossibility to relate alcoholism to a specific primum movens. Any way values for this disease are 1.2%;
– the mental retardation has lower rate than in general population; this appear to be strange since the population examined should be genetically predisposed to mental retardation, but could confirm the importance of an early diagnosis and a correct and timely treatment as useful preventive approach for homozygotes;
– values for suicide and suicide attempts are both 0.2%.
Then we analyzed the WAIS test data of 47 parents of PKU: mean

values of verbal quotient is 97.4, it is lower than the score in
general population but we cannot consider it a mental deficiency;
performance quotient scores 98.9: it appears less compromised than
the verbal quotient confirming the results of Thaalhammer (11); the
total I.Q. (Fig.1) has a value of 97.9: it is slightly low but it is
interesting to observe the wide range of values and the evidence
that a considerable part of the sample (more than 50%) is below the
value of 100.

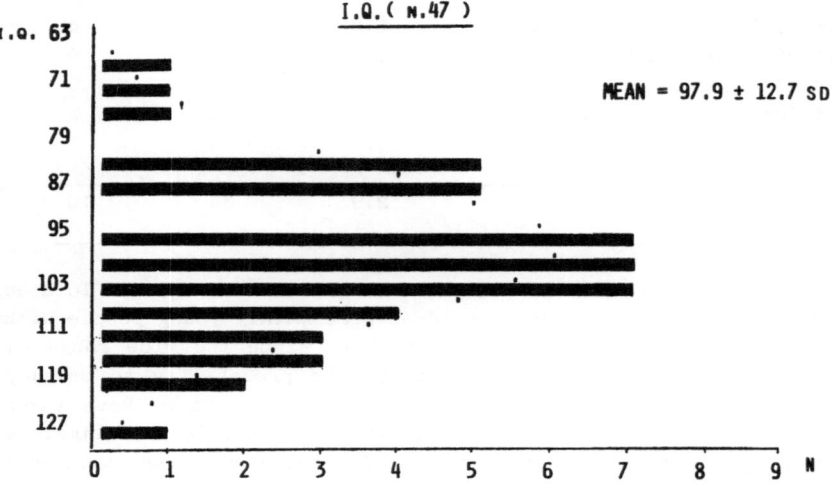

FIGURE 1. Frequencies of I.Q. values in the heterozygotes for PKU.

WAIS test appeared to be an useful tool in obtaining a good general
comprehension of the matter, since it takes into account all
intellective components of the subject examined. Beside WAIS test we
need to use for the future integrative test to better evaluate the
verbal defects, overall.

    Turning back to consider psychiatric illnesses in relatives, the
occurrence of the diseases in this group appears to be differently
distributed, and a higher frequency of affected relatives in some
families does exist. Moreover among all diagnoses we found in the
relatives schizophrenia have a rate of 22% while affective disorders
14%. The high rate shown by schizophrenic spectrum disorders could
suggest some considerations: the heterozygous state for PKU might
not to quantitatively affect the susceptibility for psychotic
disorders, while it might be a factor of interest in increasing the
development of specific psychopathological characteristics, such as
Schizophrenic Spectrum Disorders and in reducing the frequence of

other disorders; this hypothesis seems to agree with the relationship between schizophrenia and dismetabolic defects previously considered; on the other hand, the increased clustering of psychiatric illnesses within some families can be related to something more than a pure heterozygosity. In other words, we could suppose the existence of subject who have an heterozygous state for multiple amino acidopathies and therefore a more complex genetic load, not easily detectable from the phenotypical level point.

REFERENCES

1. Guttler, F (1980). Hyperphenylalaninemia: Diagnosis and classification of the various types of phenylalanine hydroxylase deficiency in childhood. Acta Paediatr Scand, suppl 280, 1.

2. Friedman, E (1969). The "autistic syndrome" and phenylketonuria. Schizophrenia, 1, 249.

3. Smeraldi, E, Lucca, A, Macciardi, F and Bellodi, L (1987). Increased concentrations of various amino acids in Schizophrenic patients: evidence for Heterozygosity effects? Hum. Genet. 38, 285

4. Blumenthal, MD (1967). Mental illness in parents of phenylketonuric children. J. Psychiat. Res. 5, 59.

5. Kang, ES, Kaufman, S, Park, SG (1970). Clinical and biochemical observations of patients with atypical phenylketonuria. Pediatrics, 45, 83.

6. Kaufman, S and Max, EE (1975). Phenylalanine activity in liver biopsies from heterozygotes: deviation from proportionality with gene dosage. Pediatr. Res. 9, 632.

7. Huether, G, Kaus, R and Neuhoff, V (1982). Brain development in experimental hyperphenylalaninemia: Mielination. Neuropediatrics, 13, 177.

8. Bauman, ML and Kemper, TL (1982). Morphologic and histoanatomic observations of the brain in untreated human phenylketonuria. Acta Neuropathol (Berl), 58, 55.

9. Giovannini, M, Riva, E, Stival, G (1979). Osservazioni sulla validità di un test di screening degli eterozigoti per la fenilchetonuria. Min. Ped. 31.

10. Diagnostic and Statistical Manual of Mental Disorders (Third Edition Revised). American Psychiatric Association, Washington D.C. 1987.

11. Thalhammer, O, Havelec, L, Knoll, E and Wehle, E (1977). Intellectual level (IQ) in heterozygotes for phenylketonuria (PKU). Hum. Genet. 38, 285.

SECTION 5:
PLASTICITY AND
VULNERABILITY OF THE CNS IN
PSYCHIATRIC DISORDERS

18
Neuronal plasticity in the central nervous system: a pharmacological approach

A. Consolazione

INTRODUCTION

In 1914 Ramon y Cajal [1] stated that "... in the adult central nervous system (CNS) the nervous pathways are something fixed, finished and immutable. All may die, nothing may be reborn". This concept aside, Cajal however pointed out that mature central fibres fail to regrow because of the absence of some "auxliary factors" or "catalytic substances", which are present during ontogenic development, but may be missing in the adult stage.

Today, in view of the most recent advances in the field of neuronal plasticity, we can consider Ramon y Cajal a pioneer not only of concepts but also of methodological perspectives. In particular, CNS lesion and transplantation studies have shown that lesioned axons possess the ability to regrow after introduction of growth-promoting signals and substrates into their environment [2,3]. Following partial deafferentation, unlesioned axons can express an inherent sprouting ability [4]. It is possible to have recovery of function as well by circumventing the need for structural repair [5]. These lines of evidence taken together point out that the mature CNS is not static, but is rather a plastic entity undergoing fine morphological and/or functional adjustments in response to perturbations in the humoral or cellular microenvironment.

The degree of a CNS "plastic" response is determined by signals derived from the humoral microenvironment, the denervated target areas and other cell-cell contacts. Among such signals, much emphasis today is being focused on neuronotrophic factors (NTFs) and neuronotoxic agents. The concentration of NTFs has been reported to increase in the adult CNS following injury [6], indicating the involvement of these factors not only in normal development, but also in the adult for survival and adequate repair processes [7]. In fact, a deficit of NTFs in the adult CNS has been proposed as a possible cause for at least some neurodegenerative diseases [8].

As concerns neuronotoxic agents, there is evidence today suggesting that excitatory amino acids, massively released in some pathological situations, are the cause of secondary death of neurons receiving an excitatory input [9]. As such, these compounds may play a key role in determining the inherent ability of brain repair processes to operate successfully.

Based on the above considerations, novel therapeutic strategies may be designed which are specifically targeted to the mechanisms underlying neuronal plasticity.

Among the pharmacological agents capable of affecting the events underlying neuronal plasticity, the monosialoganglioside GM1 (Fig. 1) is today receiving increasing attention, due to the evidence indicating that it may play a prominent role in the processes of neuritogenesis and neurite repair, both <u>in vitro</u> and <u>in vivo</u>.

The present chapter will briefly summarize the information already available regarding the potential therapeutic efficacy of GM1 in reducing damage or facilitating restoration processes in neurologically damaged subjects after head injury, stroke and perhaps neurodegenerative diseases.

GM$_1$

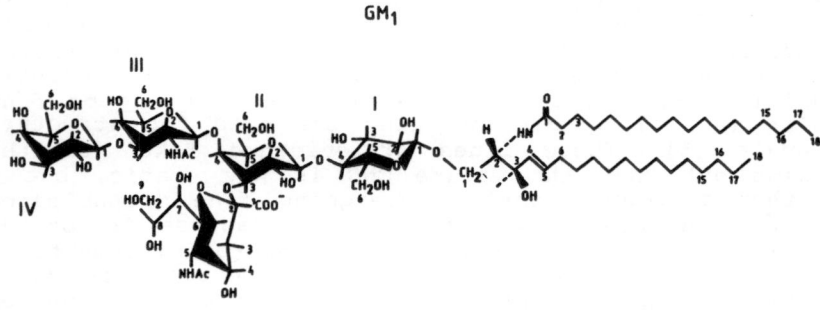

FIGURE 1 Chemical structure of GM1 monosialoganglioside

GM1 AND CNS NEURONAL PLASTICITY

Gangliosides, sialic acid-containing glycosphingolipids have long been known as characteristic surface membrane constituents of mammalian cells, in particular cells of neuronal origin. Although these molecules have been identified as receptors for bacterial toxins and viruses, relatively little is known about their function in membranes (for a review see Ledeen et al. [14]). Several lines of evidence are now available linking gangliosides, in particular GM1, to plasticity of developing and mature CNS neurons. In particular, Willinger and Schachner [11] have described changes in GM1 expression with changes in the pattern of neuronal differentiation of CNS tissue. This finding has led to postulate that GM1 may play a specific role in axon sprouting and growth. This conclusion is consistent with reports indicating that anti-GM1 antibodies produce long-lasting morphological and behavioural abnormalities, when administered to developing animals [12] and inhibit neurite regeneration both in vitro and in vivo. In addition, exogenously supplied GM1 has been reported to affect neurite outgrowth from a variety of clonal and primary (including CNS) neuronal cell systems in culture (for a review see Haber and Gorio, [13]). Many of these in vitro neuritogenic effects are associated with enhanced neuronal cell survival and/or development. However, two main features are now known to be essential for the occurrence of these GM1 effects. The first is the stable association of the exogenously added GM1 with the neuronal cell membrane; the second is the concomitant presence in the culture medium of an appropriate and adequate trophic influence (e.g. nerve growth factor, NGF) [14-18]. In fact GM1, although unable to substitute for trophic molecules, potentiates the efficacy of different trophic signals acting on specific neuronal cell types.

In addition, recent evidence shows that exogenously supplied gangliosides are also able to decrease effects of potential neuronotoxic agents, such as glutamate, in cultured neuronal cells [19].

Thus, exogenously supplied gangliosides, at least in vitro, may express dual effects: they are able to potentiate neuronal cell responsiveness to NTFs and to decrease neuronal cell responsiveness to neuronotoxic agents, in both cases resulting in an increased cell survival.

GM1 AND CNS REPAIR PROCESSES FOLLOWING DAMAGE

Based on the above observations, GM1 administration has been tested for its ability to affect CNS neuronal repair processes in adult mammmals. The systemic administration of GM1 has thus been reported to be effective in ameliorating outcome, both temporally and qualitatively, following several types of experimentally induced brain damage to adult rodents. For example, following partial unilateral hemitransection of the nigrostriatal pathway in rats, the repeated administration of GM1 was found to facilitate recovery of dopaminergic synaptic function in the lesioned striatum [20,21]. This effect was associated with both enhanced survival of nigral dopaminergic cell bodies and reduction in correlated behavioural deficits [22-24]. Analogously, GM1 was shown to facilitate maintenance of cell viability and recovery of cholinergic function following various types of lesions of the basal forebrain nuclei [25-28]. It has also been reported that chronic administration of GM1 not only restores striatal dopamine content in MPTP treated mice [29] but is capable of partially counteracting the degenerative effects on the serotoninergic and noradrenergic systems induced by 5,7-dihydroxytryptamine and 6-hydroxydopamine, respectively [30,31]. In the above mentioned paradigms, the effects of chronic GM1 treatment were assessed in the late phases following the primary insult. However, recent evidence suggests that GM1 is also effective in the "early" phases following various CNS insults, including ischemia. In particular, monosialoganglioside treatment is efficacious in protecting the brain against various biochemical and functional derangements occurring at early times following transient bilateral common carotid ligation in rats [32]. These findings corroborate those reported following cerebral ischemia in cats [33] and gerbils [34], whereby various GM1-induced biochemical and functional improvements are associated with decreased histological brain damage and reduction in mortality rate.

From the above mentioned findings it can be concluded that GM1 facilitates recovery of function by primarily decreasing neurodegenerative processes, likely by acting on those events which affect the neuronal adaptative response consequent to the occurrence of environmentally adverse conditions.

CONCLUSION

Current evidence suggests that progressive neuronal cell death following brain damage may be related to deficits in the availability of NTFs and/or increases in the presence of neuronotoxic agents, events in opposite directions which can affect the inherent capability of a plastic response by neuronal tissue. It is only relatively recently that the presence of NTFs in the adult brain [2], and their essential role have been recognized in the regulation of neuronal cell survival and repair after damage. It has been hypothesized that a neuronotrophic deficit, and concomitantly an increased trophic need occur following brain injury. Although increases in neuronotrophic activity have been reported after brain lesion [6], these increases are likely insufficient or occur at a time subsequent to most neuronal cell death. Consequently, and in accord with in vitro studies, a current working hypothesis is that the GM1-induced facilitation of CNS functional recovery in vivo is related to a facilitation of neuronal responses to trophic signals, per se inadequate at early post-lesion times. This hypothesis is further supported by recent in vivo studies whereby GM1 treatment is shown to potentiate NGF effects both in the peripheral and central nervous systems [35,36].

At the level of neuronotoxic agents, several recent investigations have provided evidence suggesting a possible imbalance between excitation and inhibition of neurons as a factor in the development of neuronal cell death. In this context, excitatory amino acid transmitters have recently been suggested to mediate brain damage due to a primary ischemic insult [9,37]. If so, in vitro GM1 inhibition of glutamate-induced translocation of protein kinase C [19] may be hypothesized to be involved in the GM1 induced prevention of secondary CNS neuronal damage in vivo.

In conclusion, the information available at present indicates that GM1 may be regarded as a novel and promising pharmacological agent, as applied to reduce damage and/or to facilitate restorative functional events in patients affected by acute, and perhaps also chronic, neurodegenerative pathologies.

REFERENCES

1.   Cajal, S Ramon y (1914). Estudios sobre la Degeneracion y Regeneracion del Sistema Nervioso. N. Moya, Madrid.
2.   Varon, S, Manthorpe, M and Williams, LR (1984). Neuronotrophic and neurite-promoting factors and their clinical potentials. Dev Neurosci, 6, 73
3.   Bray, GM, Vidal-Sanz, M and Aguayo, AY (1987). Regeneration of axons from the central nervous system of adult rats. Progr Brain Res, 71, 373
4.   Cotman, CW and Nieto-Sampedro, M (1984). Cell biology of synaptic plasticity. Science, 225, 1287
5.   Magistretti, PJ, Morrison, JH and Bloom, FE (1984). "Nervous System Development and Repair" Discussions in Neurosciences, Vol. 1(2)
6.   Nieto-Sampedro, M, Lewis, ER, Cotman, CW, Manthorpe, M, Skaper, SD, Barbin, G, Longo, FM and Varon, S (1982). Brain injury causes a time dependent

increase in neuronotrophic activity at the lesion site. Science, 217, 860

7.   Whittemore, SR, Nieto-Sampedro, M, Needels, DL and Cotman, CW (1985). Neuronotrophic factors for mammalian brain neurons: injury induction in neonatal, adult and aged rat brain. Devl Brain Res, 20, 169
8.   Appel, SH (1981). A unifying hypothesis for the cause of amyotrophic lateral sclerosis, parkinsonism and Alzheimer's disease. Ann Neurol, 10, 499
9.   Rothman, SM and Olney, JW (1986). Glutamate and the pathophysiology of hypoxic-ischemic brain damage. Ann Neurol, 19, 105
10.   Ledeen, RW, Yu, RK, Rapport, MM and Suzuki, K (1984). "Ganglioside structure, function and biomedical potential". (New York: Plenum Press)
11.   Willinger, M and Schachner, M (1980). GM1 ganglioside as a marker for neuronal differentiation in mouse cerebellum. Devl Biol, 74, 101
12.   Kasarskis, EJ, Karpiak, SE, Rapport, MM, Yu, RK and Bass, NH (1981). Abnormal maturation of cerebral cortex and behavioral deficit in adult rats after neonatal administration of antibodies to gangliosides. Devl Brain Res, 1, 25
13.   Haber, B and Gorio, A (1985). "Neurobiology of Gangliosides". (New York: Alan R Liss, Inc)
14.   Ferrari, G, Fabris, M and Gorio, A (1983). Gangliosides enhance neurite outgrowth in PC12 cells. Devl Brain Res, 8, 215
15.   Leon, A, Benvegnù, D, Dal Toso, R, Presti, D, Facci, L, Giorgi, O and Toffano, G (1984). Dorsal root ganglia and nerve growth factor: a model for understanding the mechanism of GM1 effects on neuronal repair. J Neurosci Res, 12, 277
16.   Facci, L, Leon, A, Toffano, G, Sonnino, S, Ghidoni, R and Tettamanti, G (1984). Promotion of neuritogenesis in mouse neuroblastoma cells by exogenous ganglioside. Relationship between the effect and the cell association of ganglioside GM1. J Neurochem, 42, 299
17.   Skaper, SD, Katoh-Semba, R and Varon, S (1985). GM1 ganglioside accellerates neurite outgrowth from primary peripheral and central neurons under selective culture conditions. Devl Brain Res, 23, 19
18.   Doherty, P, Dickson, JG, Flanigan, TP and Walsh, FS (1985). Ganglioside GM1 does not initiate but enhances neurite regeneration of nerve growth factor-dependent sensory neurones. J Neurochem, 44, 1259
19.   Vaccarino, F, Guidotti, A and Costa, E (1987). Ganglioside inhibition of glutamate-mediated protein kinase C translocation in primary cultures of cerebellar neurons. PNAS, 84, 8707
20.   Toffano, G, Savoini, G, Moroni, F, Lombardi, G, Calzà, L and Agnati, LF (1983). GM1 ganglioside stimulates the regeneration of dopaminergic neurons in the

central nervous system. Brain Res, 261, 163

21. Toffano, G, Agnati, LF, Fuxe, K, Aldinio, C, Consolazione, A, Valenti, G and Savoini, G (1984). Effect of GM1 ganglioside treatment on the recovery of dopaminergic nigro-striatal neurons after different types of lesion. Acta Physiol Scand, 122, 313

22. Agnati, L, Fuxe, K, Calzà, L, Benfenati, F, Cavicchioli, L, Toffano, G and Goldstein, M (1983). Ganglioside increase the survival of lesioned nigral dopamine neurons and favour the recovery of dopaminergic synaptic function in striatum of rats by collateral sprouting. Acta Physiol Scand, 119, 347

23. Sabel, BA, Dunbar, GL, Butler, WM and Stein, DG (1985). GM1 ganglioside stimulates neuronal reorganization and reduces rotational asymmetry after hemitransection of the nigro-striatal pathway. Exp Brain Res, 60, 27

24. Li, YS, Mahadik, SP, Rapport, MM and Karpiak, SE (1986). Acute effects of GM1 ganglioside: reduction in both behavioral asimmetry and loss of Na^+-K^+ ATPase after nigrostriatal transection. Brain Res, 377, 292

25. Oderfeld-Nowak, B, Skup, M, Ulas, J, Jezierska, M, Gradkowska, M and Zaremba, M (1984). Effect of GM1 ganglioside treatment on postlesion responses of cholinergic enzymes in rat hippocampus after various partial deafferentations. J Neurosci Res, 12, 409

26. Casamenti, F, Bracco, L, Bartolini, L and Pepeu, G (1985). Effects of ganglioside treatment in rats with a lesion of the cholinergic forebrain nuclei. Brain Res, 338, 45

27. Cuello, AC, Stephen, PH and Tagari, PC (1986). Retrograde changes in the nucleus basalis of the rat, caused by cortical damage, are prevented by exogenous ganglioside GM1. Brain Res, 376, 373.

28. Sofroniew, MV, Pearson, RCA, Cuello, AC, Tagari, PC and Stephens PH (1986). Parenterally administered GM1 ganglioside prevents retrograde degeneration of cholinergic cells of the rat basal forebrain. Brain Res, 398, 393

29. Hadjiconstantinou, M, Rossetti, ZL, Paxton, RC and Neff, NH (1986). Administration of GM1 ganglioside restores the dopamine content in striatum after chronic treatment with MPTP. Neuropharmacology, 25, 1075

30. Jonsson, G, Gorio, A, Hallman, H, Janigro, D, Kojima, H, Luthman, J and Zanoni, R (1984). Effects of GM1 ganglioside on developing and mature serotonin and noradrenaline neurons lesioned by selective neurotoxins. J Neurosci Res, 12, 459

31. Fusco, M, Donà, M, Tessari, F, Hallman, H, Jonsson, G and Gorio, A (1986). GM1 ganglioside counteracts selective neurotoxin-induced lesion of developing serotonin neurons in rat spinal cord. J Neurosci Res, 15, 467

32. Cahn, J, Borzeix, MG and Toffano G (1986). Effect

of GM1 ganglioside and of its inner ester derivative in a model of transient cerebral ischemia in the rat. In: Tettamanti, G, Ledeen, RW, Sandhoff, K, Nagai, Y and Toffano, G (eds.) "Gangliosides and Neuronal Plasticity". p. 435. (Padova: Liviana Press)

33.   Tanaka, K, Dora, E, Urbanics, R, Greenberg, JH, Toffano, G and Reivich, M (1986). Effect of the ganglioside GM1 on cerebral metabolism, microcirculation, recovery kinetics of ECoG and histology, during the recovery period following focal ischemia in cats. Stroke, 17, 1170

34.   Karpiak, SE, Li, YS and Mahadik, SP (1987). Gangliosides (GM1 and AGF2) reduce mortality due to ischemia: protection of membrane function. Stroke 18, 184

35.   Cuello, AC, Maysinger, D, Garofalo, L, Tagari, PC, Stephen, PH, Pioro, E and Piotte, M (1987). Influence of gangliosides and nerve growth factor on the plasticity of forebrain cholinergic neurons. In: Fuxe, K and Agnati, LF (eds.) "Receptor-receptor Interactions". p. 62 (England: Mac Millan Press Ltd)

36.   Vantini, G, Fusco, M, Bigon, E and Leon, A. GM1 ganglioside potentiates the effect of nerve growth factors in preventing vinblastine-induced sympathectomy in newborn rats. Brain Res, In Press

37.   Wieloch, T, Lindvall, O, Blomqvist, P and Gage, FH (1985). Evidence for amelioration of ischemic neuronal damage in the hippocampal formation by lesions of the perforant path. Neurol Res, 7, 24

19

A neurodevelopmental perspective on some epiphenomena of schizophrenia

R.M. Murray, M.J. Owen, R. Goodman and S.W. Lewis

Although both Kraepelin and Eugene Bleuler believed that schizophrenia was usually a deteriorating disorder, influential voices now question this. For example, Manfred Bleuler[1] estimated that 25% of his cases recovered completely, 50% ran a fluctuating course, while only 10% showed severe and permanent impairment. Several other follow-up studies have shown a more benign outcome than the traditional view would have predicted[2-4]. This has lead Zubin[5] to dispute the existence of any underlying disease process, and to postulate instead a 'vulnerability model' in which the schizophrenic individual is seen as characterised by some permanent predisposition to episodes of illness.

When enlarged cerebral ventricles and cortical sulcal widening were first demonstrated in a proportion of chronic schizophrenics, these were thought to represent the end stage of a progressive disease[6-7]. However, no correlation has been demonstrated between these changes and length of illness or treatment; indeed, they are present in young patients and soon after the onset of frank illness[8-9]. Furthermore, four groups[10-13] have now re-scanned schizophrenic patients after periods ranging up to 9 years, and have shown no significant deterioration in the CT appearances. These findings raise the question of whether the neuroradiological abnormalities could be a reflection of one of Zubin's permanent predisposing factors.

THE TIMING OF CEREBRAL INSULT

There are occasional reports of schizophrenia-like illnesses occurring as a long-term consequence of brain injury in adult life[14]. O'Callaghan et al[15] have also reported a case of schizophrenia with ventricular enlargement in whom the onset occurred two years after left frontoparietal injury at age 14. Wilcox and Nasrallah[16] recently noted that previous head injury was more common among 200 schizophrenics than in 325 patients with mania and depression, and in 134 normal subjects. Furthermore, among the schizophrenics, almost all the injuries were received before 10 years, and indeed two-thirds in infancy (before age 3 years). These findings echo earlier reports of a similar age-dependent effect of head injury as a risk factor for schizophrenia[17].

A growing number of studies have shown that obstetric complications (OC's) increase the risk of the later development of schizophrenia[18-19]. A history of OC's is found particularly in those schizophrenics with an early onset of illness and no family history of major psychiatric disorder[20-21]. CT studies of schizophrenic adults show that OC's and other early developmental hazards predict later increased ventricular size[19-22], particularly in association with widening of the cortical sulci and fissures[20].

Most studies have found a history of perinatal abnormalities to be more common than pre-natal abnormalities, but this could be because the former are easier to detect. Furthermore, some of the perinatal difficulties which were detected could have been secondary to earlier foetal damage. Schizophrenics are known to have an excess of minor physical anomalies of the type which are associated with insults to the foetus (e.g. curved fingers, furrowed tongue, low seated ears), and these anomalies are found particularly in cases with poorer premorbid adjustment and neurological impairment[23] and early onset[24].

Some[25] but not all[26] studies of monozygotic twin pairs discordant for schizophrenia, have shown the twin destined to become schizophrenic to be of lower birth weight. The issue is important since birth weight differences in MZ twins must wholly be the result of prenatal environmental factors acting differentially on the twin foetuses (e.g. twin to twin transfusion). Lewis et al[27] therefore, studied birthweight in a sub-sample of psychotic monozygotic (MZ) twins from the Maudsley twin register. Although the pre-schizophrenic twin in the discordant pairs was not always lighter, the mean difference in birth weight between the twins in these discordant pairs was more than four times greater than the intrapair difference in pairs concordant for schizophrenia. Moreover, more low birthweights (less than 2000g) were found in the discordant pairs compared to the concordant pairs. This was interpreted as evidence for prenatal environmental factors being important in discordant MZ pairs; a conclusion supported by the lower prevalence of family history of major psychiatric disorder observed in these pairs compared to the concordant pairs. The possibility that adverse prenatal factors are more common in cases of schizophrenia without obvious genetic predisposition is also supported by McNeill's[28] finding that non-familial schizophrenic patients had smaller head sizes at birth than their familial counterparts.

We have pointed out elsehwere[19-22] that the neonatal, particularly the premature, brain is very sensitive to hypoxic damage during labour. The radiological sequealae of such damage includes non-progressive enlargement of cerebral ventricles and sulci[29-30]. Hypoxic-ischaemic damage can also arise earlier. Volpe[31] who studied a series of neonates with such damage stressed the importance of asphyxia during pregnancy secondary to placental insufficiency, maternal disease or antepartum haemorrhage. Thus, it seems that a variety of complications in pregnancy and at birth can produce radiological appearances reminiscent of those reported in schizophrenia. Furthermore, although there are occasional reports of cerebral damage in adolescence or even adult life preceding schizophrenia, it is more common to obtain a history of potential cerebral insult in infancy or earlier.

Why should this be so? A possible explanation has been provided by new ideas of how the brain normally develops. It has long been held to be axiomatic that injury to the immature brain has less severe long-term consequences than does equivalent injury to the mature brain. Lately, however, animal studies have shown that there are many exceptions to this rule, and that the neuronal effects of early brain injury are complex. Although the immature brain does have greater powers of regeneration after injury than the mature brain, this regeneration may produce synaptic missconnections[32-34]. Even in adults, sprouting from transected axons can result in new connections, while axons in areas adjacent to necrosis can produce collateral sprouts which terminate on the vacated synaptic sites. When the brain is injured before the developing axons have reached their normal targets, then some of those axons will proceed to alternative destinations instead. Furthermore, the foetal brain consists of a large excess of neurones with very many interconnections not found in the normal adult brain. During development, a large proportion of these neurones die and the axons of the remainder thin out to leave a functioning 'skeleton' of neuronal pathways. Early brain damage can slow or even prevent these substractive processes causing immature patterns of cells and their projections to persist in an otherwise mature nervous system.

To our knowledge the first person to suggest that such early anomalous neuronal connections might play a role in schizophrenia was Randall[33] who wrote "input that would usually be destined for a particular neural substrate would become relayed by abnormally connected fibres into another area with which the integration of function did not 'normally' occur. Since the recipient neural substrate would not usually receive and synthesise the input that would be connected in such a manner, the output result from the different area would appear as an aberrant and extraneous variation in the function it subserves." In addition to generating aberrant responses, misconnections could also reduce general cognitive efficiency[34]. If widespread misconnections channel irrelevant information ('noise') between regions of the brain that are not normally connected, this might well interfere with the efficient functioning of each region. A poor signal:noise ratio could plausibly contribute to inattention and impairments in problem-solving ability (as reflected in lower mean IQ) in schizophrenia.

SUPPORTING EVIDENCE

Several recent post-mortem studies of patients with schizophrenia have not only shown decreased volume of temporal lobe structures[35-36] but also cytoarchitectural abnormalities which are compatible with early developmental damage or deviance. Thus, Kovelman and Scheibel[37] found heterotopic neurones and alterations in the pyramidal cell orientation particularly in the anterior and middle hippocampal regions, while Jacob and Beckman[38] found fewer neurones in certain layers throughout the cortex, and missplaced and irregularly arranged cells in the parahippocampal gyrus. Both these groups, as well as Bogerts[39] consider that such findings are highly suggestive of damage in the foetal or neonatal period. The absence of reactive gliosis also implicates damage incurred very early in life[40].

Further evidence comes from studies of epileptic patients who have undergone temporal lobectomy. While only 5% of those epileptics with mesial temporal sclerosis had concurrent psychosis, 23% of those with temporal lobe 'hamartomas' did so[41]. Taylor[41] believed that the reason for the difference was the greater likelihood of the epilepsy secondary to hamartomas beginning in adolescence, but Goodman[42] considers that the time of origin of the lesion is crucial. 'Hamartomas' arise very early in development and are already present when axonal connections are first forming, and throughout the phase of selective axonal loss. By contrast, episodes of prolonged status epilepticus causing mesial temporal sclerosis occur significantly later in brain development, probably postdating the phase when axonal connections first form. As a result of these differences, 'Hamartomas' may be more likely than mesial temporal sclerosis to initiate widespread misconnections within the limbic system. In addition, 'Hamartomas' represent a type of 'architectual folly' that appear, in some instances to be an exaggerated form of the developmental abnormalities we have described earlier; the ganglioglial type, in particular, contain wildly abnormal nerve cells. Paradoxically, both Bruton[43] and Stevens[44] have shown that the removal of 'hamartomas' by temporal lobectomy may actually induce psychosis.   This could be a consequence of leaving similar aberrant lesions in the remaining temporal lobe, or alternatively it could be secondary to terminal regeneration and aberrant sprouting in projection sites.

A number of rare developmental anomalies have also been reported in neuroradiological studies of schizophrenic patients.

these include aqueduct stenosis[45], porencephalic and arachnoid cysts[46-47]. Lewis and Mezey[48] have also drawn attention to the increased frequency among psychotic patients of cavities of the septum pellucidum. Finally, both Lewis et al[49] and Andreason[50] have reported cases of agenesis of the corpus callosum among relatively small series of schizophrenic patients although its frequency in the normal population is much lower[51].

Those who have reported such anomalies have generally interpreted the relationship with schizophrenia in straightforward neuropsychological terms on the basis of the supposed function of the malformed area of the brain. This may well be valid particularly in the defects of the corpus callosum and septum pellucidum. But, according to Barth[52] these are manifestations of neuronal migration disorders, and are usually accompanied by pervasive neuronal abnormalities. Thus, these malformations may be markers of more widespread abnormalities of neuronal development. Barth points to a wide variety of foetal insults, and specific genetic disorders, which may be causal.

We have suggested elsewhere[53] that neural dysplasia consequent upon pre-or perinatal damage results in the Type II syndrome of schizophrenia with premorbid cognitive deficits and abnormal personality as well as negative symptoms and soft neurological signs features which auger a poor prognosis. But why should disrupted neuronal migration and early misconnections produce the positive (Type I) syndrome of schizophrenia some two decades later?

The delayed onset of symptoms after early brain damage is recognised to occur when the damage affects a late-maturing region of the brain. The existence of this sort of 'sleeper' effect is well illustrated by the effect in monkeys of prefrontal lesions on delayed-response learning[54]. When the dorsolateral prefrontal cortex (DPFC) is lesioned in infancy, there is little immediate impairment on delayed-response tasks, probably because the DPFC normally plays little part in delayed-response learning during infancy. It is only in adulthood, when the DPFC generally takes over delayed-response learning, that the lesioned monkeys manifest an impairment on delayed-response tasks. In much the same way, the early-acquired deficits underlying the positive syndrome of schizophrenia might lie dormant until unmasked by the normal processes of brain maturation. Randall[33] has suggested that myelination, which continues until well into adulthood, plays a central role in unmasking the latent brain deficits in schizophrenia. This idea has subsequently been taken up by Weinberger[55]. Alternatively, it is possible that the underlying brain deficit in schizophrenia is unmasked by the progressive elimination of unwanted neural connections – a process that also continues through puberty at least. Thus in the human frontal cortex, synaptic density reaches a peak in the second year of life and then progressively declines until late adolescence, presumably as a result of selective loss[56].

SEX DIFFERENCES IN AGE OF ONSET

Why do males with schizophrenia have earlier onset of symptoms and earlier psychiatric hospitalisation than females? In England and Wales, males account for two thirds of admissions of schizophrenia aged under 35 years[57]. This sex difference in age at onset is well replicated[58-59] and does not appear to be due to artefacts such as earlier identification, or earlier presentation in males. As Rosenthal[60] points out "The male rate accelerates postpubertally and peaks in the early twenties, in accord with the original views of dementia praecox, whereas for females, the rate climbs more slowly, does not peak until the thirties, and continues to exceed that for males in later life."

There are other ways in which schizophrenia in the male more closely approximates to Kraepelin's original concept of dementia praecox than in the female. Aylward et al[61] who reviewed the evidence of intellectual impairment in schizophrenics commented that "premorbid I.Q. deficit may be an exclusive, or more pronounced characteristic of schizophrenic males". Male schizophrenics show poorer premorbid social competence[62], have a poorer response to antipsychotics, and a worse outcome than their female counterparts[63]. Males also tend to show more affective flattening while females have more florid delusions[64]. Indeed, Lewine et al[65] have shown that the stricter the diagnostic criteria of schizophrenia used, the greater becomes the ratio of male to female cases.

There is no evidence that earlier presentation of males is the consequence of genetic effect. Indeed, Kringlen[66] notes that most twin studies have found concordance rates for schizophrenia to be higher in female than male MZ twins, while Rosenthal[67] estimates the discordance rate to be twice as high in male as in female MZ pairs. This suggests a greater environmental effect in males than females, and in their prophetic twin study Rosanoff et al[68] stated that "the proportion of cases with a probable traumatic or infectious etiology seems to be higher in the male that the female sex. The most prominent factor here seems to be cerebral birth trauma." Schulsinger et al[69] also claim that the link between obstetric complications and schizophrenia is stronger in males than females. Since schizophrenics with a history of obstetric complications are known to present earlier than other cases, this may offer an explanation of the earlier onset in males.

Why should obstetric complications be more important in the etiology of male than female schizophrenia? Two possibilities come to mind. Firstly, Amato et al[70] have shown that peri- and intraventricular haemorrhage are more common in male than female premature infants even when there are no differences between the sexes in birth weight, gestational age or major perinatal complications. These authors concluded that this greater vulnerability is due to the slower maturation of the cerebral vasculature in the male. Rantakallio and Vont Wendt[71] have also shown that among low birthweight infants followed up to age 14, boys have more educational and neurological problems than girls.

Secondly, the functions of the cerebral hemispheres are less strongly lateralised in females who therefore have more bilateral representation of functions such as language. Flor-Henry[72] suggests that an insult to the left hemisphere is, therefore, likely to be more crucial in males than females. He points out that developmental disorders involving speech such as infantile autism, childhood dyslexia and stutering are all at least 4-5 times more common in males than females.

ABNORMALITIES OF HANDEDNESS

Flor-Henry[73] also showed that epileptics who develop schizophrenia-like psychoses tend to have foci in the dominant temporal lobe. Since then many other workers have looked at direct and indirect measures of hemispheric structure and function in schizophrenia, the majority implicating abnormality of the left rather than the right hemisophere[74]. Some but not all[74] studies have also suggested an increased prevalence of left-handedness among schizophrenic patients. Could the excess occur only in certain types of patients? Lishman and McMeekan[75] reported that young male psychotic patients had a higher than expected proportion of left-handers (over 25% compared to 10% in other psychiatric patients). Hauser et al[76] noted that among female schizophrenics, non-right handedness was particularly associated with long labour and with early onset of psychosis.

These results raise the question of whether left-handedness is more common only among those schizophrenics who have been subject to neurodevelopmental insult. An unconfirmed survey of the relatives of over 900 schizophrenics suggests that probands who had suffered obstetric complications are more likely to be left-handed[77]. Left-handedness is increased among premature children and others in whom the process of cerebral lateraliation has been disrupted[78-79]. Satz and his colleagues[80] maintain that early left-sided cerebral lesions may induce a switch of cerebral dominance, and produce a distinct syndrome of "pathological left-handedness". Henry et al[81] support this notion by pointing out that such individuals rarely have a family history of left-handedness, and that left-handedness is 2-3 times as common in epileptics and mentally handicapped individuals as in the normal population.

If neurodevelopmental insults are associated with an increase of left-handedness among schizophrenics, then one might expect

sinistrality and indices of left-sided brain damage to be more
prevalent among non-familial schizophrenics. Both Nasrallah et
al[82] and Shimizu et al[83] reported that left-handed
schizophrenics were less likely than their right-handed
counterparts to have a family history of schizophrenia. In their
CT study of MZ twins discordant for schizophrenia, Reveley et
al[84] concluded that relative hypodensity of the left hemisphere
was environmental rather than genetic in origin. Hays[85] showed
that non-familial schizophrenics were more likely than their
familial counterparts to have abnormal EEG's, and that the
abnormalities tended to be left-sided. Finally, Lishman, et
al[86] found that non-familial schizophrenics had exaggerated ear
differences on a dichotic listening test.

SEASON OF BIRTH

Many studies have shown an excess of late winter or spring births
amongst schizophrenics of between five and fifteen per cent[87].
Although there are other possible explanations, that which has
attracted most interest is the "constitutional damage" hypothesis
which holds that birth in the early months of the year is
associated with an increased risk of cerebral damage which
predisposes to later schizophrenia.

This hypothesis predicts that schizophrenics with no obvious
genetic predisposition should show more winter births than those
at high genetic risk. Kinney and Jacobson[88] studied 34
schizophrenic adoptees and found a significant excess of low
genetic risk patients born in the early months of the year.
McNeil[28] also reported a significant negative relationship
between a family history of schizophrenia and birth in January to
April among 70 schizophrenic patients. Schur[89] who studied 975
schizophrenics, noted that those with schizophrenic relatives were
born less often in the first quarter of the year by 29% compared
with those without such relatives.

Indirect support comes from studies which have examined
paranoid versus non-paranoid schizophrenics, for the former are
known to be less likely to have affected relatives than the
latter. Hsieh et al[90] were unable to show an excess of winter
births among 472 schizophrenic patients but did indeed find a
significant excess in the 102 male paranoid schizophrenics.
Similarly, Nasrallah et al[92] found that among 165 male paranoid
schizophrenics there was an excess in the cold winter months of
the year (although confusingly the opposite was the case in
females). Templar and Veleber[92] noted larger inverse
correlations between the monthly temperature and births of
paranoid schizophrenics than the catatonic-hebephrenic group.

What could the mechanism be? Many viral infections vary in
their seasonality as do their adverse neurological consequences.

For example, there are seasonal variations in the monthly births of individuals with congenital rubella[93] and febrile convulsions[94]. Watson et al[95] claimed that the winter birth excess of schizophrenia was greatest following those years marked by high levels of infectious diseases, particularly diphtheria, pneumonia and influenza, all diseases likely to increase the frequency of hypoxic damage to the foetus. Mednick et al[96] also report that influenza infection during the second trimester of pregnancy increases the risk to the foetus of later schizophrenia.

Videbech et al[97] found a correlation between the monthly indices for stillbirths and for birth of schizophrenics in Denmark, and concluded that this was compatible with a role for "intra-uterine and perinatal complications in the genesis of schizophrenia". If, perinatal complications are to blame then one would expect that those schizophrenics who as neonates were at increased risk of difficult labour should show an especially large season of birth effect. Twins are such a population, but Kendler[98] reported only a non-significant 4.1% excess of births in the first quarter of the year in 536 schizophrenic twin pairs. Similarly, McNeil[28], who found that non-familial schizophrenics were preferentially born in the winter months, was unable to find any winter excess in the monthly distribution of obstetric complications.

If the excess of winter born schizophrenics is a consequence of neurological damage to the foetus or neonate, then such patients should show greater abnormality on CT scan than other schizophrenic subjects. The difficulty, of course, is that it is impossible to distinguish the excess schizophrenics from those who would have been born in the winter months anyway. Nevertheless, two studies have examined CT scans in winter born versus summer born schizophrenics. Zidinski and Schulz[99] found that nine schizophrenics born between December and March had significantly larger cerebral ventricles than eleven schizophrenics born between April and July. It is hard to ascribe the result of this small study to anything other than chance, but Sacchetti et al (1987) carried out a similar analysis on a more substantial group of schizophrenic subjects (n=115). Thirty six per cent of those born between December and April had enlarged ventricles compared with only 16% of those born between May and November, a risk 2.7 times greater.

We attempted to replicate this finding in a sample of 124 schizophrenics. As can be seen (Table), there were no significant differences in ventricular size between those born in the winter months and in the summer months. Similarly, there was no significant difference in the distribution of obstetric complications or in familial versus non-familial cases. However, in order to address the question of whether the excess of winter-born schizophrenics have more cerebral damage one must first demonstrate a seaonslity of birth effect in the schizophrenic population under study. There was no excess of winter births in our sample, and Schur[89] has demonstrated that schizophrenic patients admitted to the Maudsley Hospital no longer show a season of birth effect. Indeed, it seems that the seaonality of a variety of disorders is showing a secular decline in recent years possibly as a result of better ante- and peri-natal care.

CONCLUSIONS

It is our contention that schizophrenia is aetiologically heterogeneous, and that the type which most closely approximates to Kraepelin's original concept of dementia praecox is consequent upon neurodevelopmental damage, most commonly in foetal or neonatal life. The resulting neural dysplasia results in many of the negative signs of schizophrenia while the positive syndrome develops when a brain containing early anomalous connections undergoes the normal morphological changes of adolescent and young adulthood. The fact that this maturational phase is time-limited explains why schizophrenics tend not to deteriorate after the first five years. A neurodevelopmental perspective may also help to explain some of the other puzzles of schizophrenia, namely the winter birth excess, the earlier onset in males, and the increase in left-handedness reported in some studies.

VENTRICULAR BRAIN RATIO, OBSTETRIC COMPLICATIONS AND FAMILY HISTORY BY MONTH OF BIRTH

MONTH OF BIRTH	Jan.	Feb.	March	April	May	June	July	Aug.	Sept.	Oct.	Nov.	Dec.
VENTRICULAR BRAIN RATIO (VBR) (n = 124)												
No. of Subjects	17	6	8	14	12	12	7	9	8	14	9	8
Mean VBR	7.15	7.44	6.15	6.71	6.58	7.60	7.82	7.57	7.02	9.07	7.26	6.29
STD DEV.	2.87	3.36	3.37	2.64	2.76	3.82	2.63	2.91	2.36	4.66	3.46	2.76
OBSTETRIC COMPLICATIONS* (n = 11)												
Absent	12	3	5	12	13	8	3	5	6	8	4	4
Equivocal	1	1	3	2	0	1	0	2	1	1	0	0
Definite	0	2	0	0	0	3	2	1	0	3	3	2

*According to the scale of Lewis et al (1988)

FAMILY HISTORY OF MAJOR MENTAL ILLNESS IN A FIRST DEGREE RELATIVE (n = 126)												
Absent	8	5	7	10	9	9	6	7	7	10	8	3
Present	9	1	2	4	4	3	1	2	1	4	1	5

1. Bleuler, M (1978). The schizophrenic disorders. New Haven Yale University Press

2. Bland, RC Parker, JH and Orn, H (1976). Prognosis in schizophrenia. Arch. Gen. Psychiat., 33, 949

3. Ciompi, L (1980). The natural history of schizophrenia in the long-term. Brit. J. Psychiat, 136, 413

4. Wing, JK (1987). Has the outcome of schizophrenia changed? Brit. Med. Bull. 43, 741.

5. Zubin, J. (1988). Chronicity versus vulnerability. In Tsuang, MT and Simpson, JC (Eds) Nosology, Epidemiology and Genetics of Schizophrenia p. 463. Elsevier Amsterdam.

6. Johnstone, EC Crow, TJ Frith CD Husband, J and Kreel, L (1976). Cerebral ventricular size and cognitive impairment in chronic schizophrenia. Lancet, 2, 924.

7. Woods, BT and Wolf, J (1983). A reconsideration of the relation of ventricular enlargement to duration of illness in schizophrenia. Am. J. Psychiat, 140, 1564.

8. Shelton, RC and Weinberger, DR (1986). CT studies in schizophrenia. In Nasrallah, H and Weinberger, DR. The Neurology of Schizophrenia, p 207. Elsevier, Amsterdam.

9. Reveley, M and Trimble MR (1987). Application of imaging techniques. Brit. Med. Bull., 43, 615.

10. Nasrallah, HA Olsen, SC McCalley-Whitters, M Chapman, S and Jacoby, GC (1986). Cerebral ventricular enlargement in schizophrenia: a preliminary follow-up study. Arch. Gen. Psychiat., 43, 157.

11. Illowsky, B Juliano, DM Bigelow, LB and Weinberger, DR (1988). Stability of CT scan findings in schizophrenia. J. Neurol. Neurosurg. and Psychiat, 51, 209.

12. Reveley, M Chitkara, B and Lewis, S (1988). Unpublished data.

13. Sacchetti, E, Vita, A Calzeroni, A Invernizzi, G and Cazullo, CL (1987). Neuromorphological correlates of schizophrenic disorders. In Cazullo et al (eds). Etiopathogenetic Hypotheses of Schizophrenia. MTP Press Lancaster.

14. Davison, K and Bagley, CR (1969). Schizophrenia-like psychoses associated with organic disorder of the CNS. R.N. Herrington (Ed). Current Problems in Neuropsychiatry. Headley, Kent.

15. O'Callaghan, E and Larkin, C (1988). Early onset schizophrenia after teenage head injury. Brit. J. Psychiat. In Press.

16. Wilcox, J.A. and Nasrallah, HA (1987). Childhood head trauma and psychosis. Psychiat. Res, 21, 303.

17. Achte, K.A., Hillbrom, E. and Aalberg, V (1969). Psychosis following war brain injuries. Acta Psychiat. Scand., 45, 1.

18. McNeill, TF and Kaij, L (1978). Obstetric factors in the development of schizophrenia. Wynne, LC, Cromwell, RL and Matthysse, S. The Nature of Schizophrenia p. 401. Wiley, New York.

19. Lewis, SW, Owen MJ and Murray, RM (1988). Obstetric complications and schizophrenia; methodology and mechanisms. Tamminga, CA and Schulz, SC. Schizophrenia: A Scientific Focus. O.U.P. New York.

20. Owen, M Lewis SW and Murray, RM (1987). Obstetric Complications and cerebral abnormality in schizophrenia. In Cazullo et al (eds) Etiopathogenetic Hypothesis of Schizophrenia. MTP Press. Lancaster.

21. O'Callaghan, E Larkin, C and Waddington, J (1988). Paper presented at the 4th Biennial Winter Workshop on Schizophrenia. Badgastein.

22. Murray, RM Lewis SW and Reveley AM (1985) Towards an aetiological classification of schizophrenia. Lancet, i, 1023.

23. Guy, JD, Majorski, LV, Wallace, CJ and Guy, MP (1983). The incidence of minor physical anomalies in adult male schizophrenics. Schiz. Bull., 9, 571.

24. Green, MF Satz, P Soper, HV and Kharabi, F (1987). Relationship between physical anomalies and age at onset of schizophrenia. Amer. J. Psychiat, 144, 666

25. Mosher, LR, Pollin, W and Stabenau, JR (1971). Identical twins discordant for schizophrenia. Arch. Gen. psychiat, 24, 422.

26. Kringlen, E (1967). Heredity and environment in the functional psychoses. London, William Heinemann.

27. Lewis, SW Chitkara, B Reveley AM and Murray, RM (1987). Family history and birthweight in MZ Twins Concordant and Discordant for psychosis. Acta Genet. Med. Gemmellol., 36, 267.

28. McNeill, TF (1988) Paper presented at the 4th Biennial Winter Workshop on Schizophrenia. Badgastein.

29. Leichty, EA Gilmore, RL Bryson, CQ and Bull, J (1983) Outcome of high risk neonates with ventriculomegaly. Devel. Med. Child. Neurol, 25, 162.

30. DeVries, LS Dubowitz, LMS Dubowitz, V et al (1985). Predictive value of cranial ultrasound in the newborn baby. Lancet,2, 137.

31. Volpe, JJ (1981) Neurology of the Newborn. WB Saunders, Philadelphia.

32. Janowsky, JS and Finlay, BL (1986) The outcome of perinatal brain damage. Devl. Med. Child Neurol., 29, 243.

33. Randall, PL (1983) Schizophrenia, abnormal connection and brain evolution. Medical Hypothesis, 10. 247.

34. Goodman, R. (1989). Limits to Cerebral Plasticity. In Johnson, DA Utley, D Wyke, M (Eds) Children's Head Injury: who cares? Falmer, Basingstoke.

35. Bogerts, B Meertz, E Schonfeld-Bausch, R (1985) Basal ganglia and limbic system pathology in schizophrenia. Arch. Gen. Psychiat. 42, 784.

36. Brown, R, Colter, N, Corsellis, J., Crow, T., Frith, CD, Jagoe, R, Johnstone, EC and Marsh, L (1986). Post-mortem evidence of structural brain changes in schizophrenia. Arch. Gen. psychiat., 43, 36.

37. Kovelman, JA and Schiebel, AB (1984) A neurohistological correlate of schizophrenia. Biol. Psychiatry, 19, 1601.

38. Jacob, H and Beckman, H (1986) Prenatal development disturbances in the limbic allocortex in schizophrenics. J. Neural. Transmiss., 65, 303.

39. Bogerts, B. (1988). Limbic and Paralimbic Pathology in Schizophrenia. In Tamminga, CA and Schulz, C. Schizophrenia: A Scientific Focus. O.U.P. New York.

40. Roberts, GW Colter, N Lofthouse, R Johnstone, EC and Crow, T (1986). Is there gliosis in schizophrenia? Biol. Psychiat. In Press.

41. Taylor, DC (1975) Factors influencing the occurrence of schizophrenia-like psychosis in patients with TLE. Psychol. Med., 5, 249.

42. Goodman, R. (1988). Neuronal missconnections and psychiatric disorder. Brit. J. Psychiat. In Press.

43. Bruton, C (1988) A Neuropathological Study of Temporal Lobe Epilepsy. Maudsley Monograph. O.U.P. In Press.

44. Stevens, J (1988) Paper presented at the 4th Biennial Winter Workshop on Schizophrenia, Badgastein.

45. Reveley, A.M. and Reveley, MA (1983). Aqueduct Stenosis and Schizophrenia. J. Neurol. Neurosurg. Psychiat., 46, 18.

46. Blackshaw, S and Bowen, RC (1987) A case of atypical psycosis associated with alexithymia and a left fronto-temporal lesion. Can. J. Psychiat., 32, 688.

47. Lewis, SW (1986) Schizophrenics with and without intracranial abnormalities on CT. MPhil Thesis. Univ. of London.

48. Lewis, SW and Mezey, GC (1985) Clinical correlates of septum pellucidum cavities. Psychol. Med., 15, 43.

49. Lewis, SW, Reveley, MA David, AS and Ron, MA (1988) Agenesis of the Corpus Callosum and Schizophrenia. Psychol. Med. In Press.

50. Andreason, N (1988) Paper presented at the 4th Biennial Winter Workshop on Schizophrenia, Badgastein.

51. Njiokiktjien, C (1988) Paedriatric Behaviour Neurology, Suyi, Amsterdam.

52. Barth, PG (1987). Disorders of Neuronal Migration. Can. J. Neurol. Sci, 14, 1.

53. Murray, RM Lewis, SW Owen, MJ and Foerster, A (1988). The Neurodevelopmental Origins of Dementia Praecox. In McGuffin, P and Bebbington, P. (Eds) Schizophrenia; the Major Issues, Heinemann, London.

54. Goldman-Rakic, PS (1987) Circuitry of the prefrontal cortex and the regulation of behaviour by representational knowledge. In Plum, F and Mountcastle, V. Handbook of Physiology vol. 5, p 373. American Physiological Society. Bethesda.

55. Weinberger, DR (1987). Implications of normal brain development for the pathogenesis of schizophrenia. Arch. Gen. Psychiat, 44, 660.

56. Huttenlocher, PR (1979). Synaptic density in human frontal cortex-developmental changes and effects of ageing. Brain Res, 163, 195.

57. Department of Health and Social Security (1986). Statistical Bulletin No. 1. London.

58. Lewine, R (1981). Sex differences in schizophrenia. Psychol. Bull., 90, 432.

59. Loranger, AW (1984). Sex difference in age of onset of schizophrenia. Arch. Gen. Psychiat., 41, 157.

60. Rosenthal, D. (1971) Genetics of Psychopathology. McGraw-Hill. New York.

61. Aylward, E Walker, E and Bettes, B (1984). Intelligence in schizophrenia. Schiz. Bull., 10, 430.

62. Kokes, R Strauss, JS and Klorman, R (1977). Measuring premorbid adjustment. Schiz. Bull, 3, 186.

63. Seeman, MV (1982). Gender differences in schizophrenia. Can. J. Psychiat, 27, 107.

64. Katschnig, H. and Lenz, G (1988). Paper presented at the 4th Biennial Winter Workshop on Schizophrenia. Badgastein.

65. Lewine, R Burbach, D and Meltzer, H (1984) Effects of diagnostic criteria on the ratio of male to female schizophrenic patients. Am. J. Psychiat. 141, 84.

66. Kringlen, E (1987) Contributions of Genetic Studies on Schizophrenia. Hafner, H Gattaz, W and Janzarik, W. The Search for the Causes of Schizophrenia p.123. Springer-Verlag. Berlin.

67. Rosenthal, D. (1962). Familial concordance by sex with respect to schizophrenia. Psychol. Bull., 59, 401.

68. Rosanoff, A Handy, L Plesset, I and Brusch, S (1934) The Etiology of so-called schizophrenic psychoses. Amer. J. Psychiat, 91, 247.

69. Schulsinger, F, Mednick, SA Walker, EF Cudeck, R and Moffitt, T (1980). Biosocial implications growing from high-risk research. Acta Psychiat. Scand, Suppl. 285, 62, 112.

70. Amato, M, Howald, H. Von Muralt, G (1987) Foetal sex distribution of peri-intraventricular haemorrhage in preterm infants. Eur. Neurol., 27, 20.

71. Ratakallio, P and Von Wendt, L. (1985). Prognosis for low birth weight infants up to the age of 14. Dev. Med. Child. Neurol., 27, 655.

72. Flor-Henry, P (1978). Gender, hemispheric specialisation and psychopathology. Soc. Sci, and Med., 12B, 155.

73. Flor-Henry, P. (1969). Psychosis and temporal lobe epilepsy. Epilepsia, 10, 363.

74. Taylor, PJ (1987) Hemispheric lateraliation and schizophrenia. In Helmchen, H and Henn, FA. Biological Perspectives of Schizophrenia p 125. John Wiley, Chichester.

74. Lishman WA and McMeekan, ERL (1976). Hand preference patterns in psychiatric patients. Br. J. Psychiat. 129, 158.

76. Hauser, P Pollock, B, Finkelberg, F, et al (1985). On sinistrality and sex differences in schizophrenia. Am. J. Psychiat. 142, 1228.

77. Tyler, M. (1988) Personal communication.

78. O'Callaghan, MJ Tudehope, DI Dugdale, AE Mohany, H Burns, Y and Cook, F (1987). Handedness in children with birthweights below 1000 gm Lancet, i, 1155.

79. Ross, G Lipper EG Auld PAM (1987) Hand preference of four year old children. Devel. Med. Child. Neurol, 29, 615.

80. Orsini, DL and Satz, P (1986). A syndrome of pathological left-handedness. Arch. Neurol, 43, 333.

81. Henry, RR Satz, P and Saslow, E (1984) Early brain damage and the ontogenesis of functional asymmetry. In Early Brain Damage Vol. 1. Academic Press. New York.

82. Nasrallah, HA McCalley-Whitters M and Kuperman, S (1982). Neurological differences between paranoid and non-paranoid schizophrenia. J. Clin. Psychiat., 42, 305.

83. Shimizu, A Yamaguchi, H and Isaki, K. Hand preference in schizophrenics and handedness conversion in children. Acta Psychiat. Scand. 72, 259.

84. Reveley, M. Reveley, A and Baldy R (1987) Left cerebral hypodensity in discordant schizophrenic twins. Arch. Gen. Psychiat, 44, 625.

85. Hays, P (1977) Electroencephalographic variants and genetic predisposition to schizophrenia. J. Neurol. Neurosurg. Psychiat., 40, 753.

86. Lishman, WA, Toone, BK and Colbourne, CJ (1978) Dichotic listening in psychotic patients. Br. J. Psychiat. 132, 333.

87. Bradbury, TN and Miller, A (1985) Season of birth in schizophrenia. Psychol. Bull., 98, 569.

88. Kinney, DK and Jacobsen, B (1978) Environmental factors in schizophrenia. Wynne, LC, Cromwell, RL and Matthysse, S. The Nature of Schizophrenia. p. 38. Wiley, New York.

89. Shur, E (1982). Season of birth in high and low genetic risk schizophrenics. Brit. J. Psychiat. <u>140</u>, 410.

90. Hsieh, H., Khan, MH Atwal, SS and Cheng, SC (1987) Season of birth and subtypes of schizophrenia. Acta Psychiat. Scand., <u>75</u>, 373.

91. Nasrallah, H and McCalley-Whitters, M (1984) Seasonality of births in subtypes of chronic schizophrenia. Acta Psychiat. Scand., <u>69</u>, 292.

92. Templer, DI and Veleber, DM (1982) Seasonality of schizophrenic births. Brit. J. Psychiat, <u>140</u> 323.

93. Peckham, CS (1978). Congenital Rubella surveillance. J. Roy. Coll. Phycns, <u>12</u>, 250.

94. Sunderland, R., Carpenter, RG and Gardner, A (1982). Are all born equal? Brit. Med. J. <u>284</u>, 624.

95. Watson, CG Kucala T Tilleskjor, C and Jacobs, L (1984) Schizophrenic birth seasonality in relation to the incidence of infectious diseases. Arch. Gen. Psychiat. <u>41</u>, 85.

96. Mednick, SA Machon, RA Huttenen, MO and Bonnett, D (1988) Adult Schizophrenia following prenatal exposure to an influenza epidemic. Arch. Gen. Psychiat, <u>45</u>, 189.

97. Videbech, TL Weeke, A and Dupont, A (1974) Endogenous psychoses and season of birth. Acta. Psychiat. Scand, <u>50</u>, 202.

98. Kendler, KS (1982) Season of birth of schizophrenic, neurotic and psychiatrically normal twins. Br. J. Psychiat. <u>141</u>, 186.

99. Zipursky, RB and Schulz, SC (1987) Seasonality of birth and CT findings in schizophrenia. Biol. Psychiat. <u>22</u>, 1288.

100. Sacchetti, E Vita, A Battaglia, M Calzeroni A Conte, T Invernizzi, G and Cazullo, CL (1987). Season of birth and cerebral ventricular enlargement in schizophrenia. In Cazullo et al (eds) Etiopathogenetic Hypothesis of Schizophrenia. MTP Press Lancaster

20
Clinical investigations of plasma homovanillic acid concentrations

M. Davidson, R. Kaminsky, S. Jaff, N. Runyon and K.L. Davis

Several lines of evidence suggest that abnormalities of central dopaminergic transmission may be involved in the expression of some schizophrenic symptoms. However, elucidation of the role of dopamine (DA) in schizophrenia has eluded investigative efforts. Presently, no accurate and easily repeatable measure of brain DA activity exists. Measurements of CSF homovanillic acid concentrations and determination of growth hormone and prolactin plasma levels, whose secretion is under central dopaminergic control, has major limitations. A more promising technique, positron emission tomography, is not yet available for routine research use.

In recent years, the development of a technique to measure homovanillic acid in plasma has offered the possibility of performing serial measurements of this major DA metabolite. Plasma homovanillic acid (pHVA) originates both from the brain and from peripheral sources. The purpose of this investigation is to:

1) attempt to provide evidence that pHVA concentrations are indeed affected by brain HVA concentrations;

2) examine confounding exogenous factors affecting measurements of pHVA;

3) examine pHVA concentrations in schizophrenic patients.

1a) Effects of haloperidol and haloperidol combined with debrisoquin on pHVA concentrations in human subjects.

The value of pHVA in clinical psychiatric research is dependent upon the degree to which it accurately reflects brain DA function. The concentration of HVA in the brain reflects the rate of DA turnover, and presumably, presynaptic neuronal activity. Thus, if HVA changes in parallel in both plasma and brain, pHVA will indeed provide an accurate measure of brain DA function.

Pharmacological probes affecting central DA transmission in rodents produce parallel changes in brain HVA and pHVA. These changes are not affected by pretreatment with debrisoquin (1,2). Debrisoquin is a monoamine oxidase (MAO) inhibitor which does not penetrate the blood-brain-barrier (3), hence it suppresses the formation of peripheral HVA (4), but not the formation of central HVA (5,3). Therefore, following debrisoquin administration, haloperidol or apomorphine induced changes in pHVA reflect central, and not peripheral effects (1).

The purpose of this study was to investigate whether we can detect in humans, like in animals, an increase in pHVA concentrations shortly after challenge with the neuroleptic drug haloperidol. Twenty-eight schizophrenic patients who were free of oral neuroleptics for at least three weeks, and depot neuroleptics for at least three months, participated in the study. pHVA was sampled between 0830 and 1700. Haloperidol (0.2 mg/kg) was administered intramuscularly at 1100, and 10mg was given orally at 1800. pHVA concentrations were sampled again the next morning at approximately 0900. To confirm that any haloperidol induced changes in the pHVA concentration arise from central DA systems and not from peripheral sources, a subgroup of ten patients were pretreated with 20mg/d of debrisoquin sulfate for 7 days before challenge with haloperidol.

Haloperidol produced increases in pHVA concentrations in all but two patients (mean pHVA concentration before haloperidol, 9.1 ± 0.9 ng/ml; after haloperidol, 12.8 ± 1.4 ng/ml; t (17) = 3.7; p .001). Pretreatment with debrisoquin produced an initial decrement in the pHVA concentration. However, debrisoquin did not diminish the magnitude of the haloperidol-induced increases in the pHVA level (before haloperidol, 4.3 ± 0.3 ng/ml; after haloperidol, 8.2 ± 1 ng/ml; t (9) = 4; p .003). Thus, the haloperidol induced changes in pHVA were of central, not peripheral origin, confirming the previous animal experiments. Results of this investigation strongly suggest that the acute effect of neuroleptics on brain dopaminergic activity can be reflected in changes in pHVA concentration. The ability of haloperidol to produce central pHVA increments in human subjects was recently replicated in an independent study (6). However, haloperidol did not produce acute pHVA elevations in all patients, nor was the magnitude of the increments equal among patients in whom the elevations occurred. Although not reaching statistical significance, it was observed by us and by Davila (6) that patients with the best treatment response were those in whom the greatest pHVA elevations were noted. It was hypothesized that this phenomenon may be reflective of the ability of the dopaminergic system to maintain its plasticity (7).

1b) Relationship between cortical atrophy and pHVA concentrations in schizophrenic patients.

Although DA concentration in the frontal cortex is relatively low, DA turnover in this brain area is high. Differences in pHVA concentration between schizophrenic patients with and without frontocortical atrophy would suggest that the contribution of the frontal cortex to the total pHVA concentration is considerable and that a drastic reduction in this contribution could be detected by measuring pHVA concentrations. Such results would offer additional support to the idea that brain HVA concentrations affect pHVA concentrations.

CT scans were performed in 28 schizophrenic patients. Cortical atrophy was determined as the average rating of two independent raters using the standard clinical scale based on subjective ratings ranging from 3-0 (normal, mild, moderate and severe). There was complete agreement between the two raters who were blind to all identifying information about the subjects (including age). The VBR did not correlate with pHVA levels, and no significant differences in pHVA were noted in comparing the thirteen subjects with the largest VBR values to the fourteen with the smallest VBR values. Fifteen subjects showed no cortical atrophy, 8 showed mild, 4 moderate, and 1 showed severe cortical atrophy. Cortical atrophy did not correlate with pHVA in this sample (Spearman r = -18, df = 26, p = NS). However, in those 5 subjects demonstrating moderate or severe cortical atrophy, there was a significantly lower mean level of baseline pHVA (7.2 + 2.2 ng/ml) as compared to the 22 subjects without CT scan evidence of serious cortical atrophy (2.2 + 5.6 ng/ml; two-tailed t-test, t = 3.22, df = 25, p .005). The group with significant cortical atrophy was older (48.4 + 8.4 years) than the comparison group (34.5 + 10.0 SD years; two-tailed t-test, t = 2.90, df = 25, p .01). However, age per se did not correlate with pHVA (Pearson r = .15, df = 25, p .4), and since there was still a significant group difference using an analysis of covariance, with age as a covariant (F (2,24) = 8.13, p .01), a hidden age effect did not appear to account for the relationship between cortical atrophy and pHVA.

The finding of lower concentrations of pHVA in schizophrenics with cortical atrophy compared to those without cortical atrophy is in agreement with another independent study (8) showing that schizophrenic patients with cortical atrophy have low pHVA concentrations which are further lowered by alprazolam administration. Low pHVA concentrations may reflect less HVA production from a diminished mass of cerebral cortex, where DA concentrations are low but turnover is high (9). Combined with the large volume of cortex normally present, the cortical mass may be sufficient to produce considerable amounts of HVA, and thus cortical atrophy would result in decreased HVA levels.

If this simple physical interpretation is correct, and if dopaminergic neurons are lost in proportion to cortical mass loss, it may prove to be vital to determine total cortical mass in subjects in whom pHVA studies are performed. Alternately, dopaminergic activity in schizophrenic individuals who suffer from cortical atrophy may differ in a fundamental way from dopaminergic activity in schizophrenics without such atrophy. Subjects with cortical atrophy may be particularly limited in their capacity to activate the dorsolateral prefrontal cortex, an area rich in dopaminergic projections, and hence have a clinical picture affected by this limitation (10).

2a) Effects of diet on pHVA concentrations in human subjects

HVA is produced in the periphery from DA precursors absorbed from the GI tract. In monkeys, pHVA concentrations were almost three times greater following a meal high in monoamine contents compared with fasting animals. This suggests that the dietary effect must be controlled in any study attempting to reflect central DA activity by measuring pHVA concentrations (11). Six subjects participated in an experiment examining monoamine content on pHVA over an 18-hour period. A third diet, a pure carbohydrate meal (polycose, Ross Laboratories) was used as a control. At weekly intervals, subjects ingested one of the three meals between 1800 and 1830. Samples for pHVA were drawn at half hour intervals through an IV line between 1600 and 2100 and again between 0800 and 1000 the next morning.

Postmeal (30 minutes to 180 minutes after food was consumed), there were significant effects due to diet ($F = 25.7$; df = 2/10; $p = 0.002$), time ($F = 4.7$; df = 5/25; $p = 0.02$), and the time by diet interaction ($F = 6.7$; df = 10/50; $p = 0.001$). Comparison of the high monoamine condition with the polycose condition indicated significant effects of diet ($p = 0.003$), time ($p = 0.01$), and a time by diet interaction ($p = 0.002$). Comparison of the low monoamine condition with the polycose condition showed an effect of time ($p = 0.05$) but no effect of diet ($p = 0.3$) or the time by diet interaction ($p = 0.26$). Thus, these results confirm an earlier study (12), indicating a dramatic rise in mean pHVA following a high monoamine diet, but not after a low monoamine diet. It was, however, of utmost importance to establish what the time period was over which dietary intake can affect pHVA concentration. Without this information, unnecessary dietary restrictions could have been imposed on patients participating in pHVA studies. Whether this dietary effect lasts overnight was examined by analyzing the day-after time points. Five time points from between 0800 and 1000 were studied. Neither diet ($F = 2.4$; df = 2/10; $p = 0.14$), nor the diet by time interaction ($F = 2.1$; df = 8/40; $p = 0.17$) was significant, though there was a marginal effect of time (($F = 3.1$; df = 4/20; $p = 0.07$).

As is evident from this study the elevation of pHVA in the high monoamine condition did not persist until the next morning; rather pHVA tended to drop slightly in all conditions during the early morning. Results of this investigation demonstrated that the effects on pHVA of a meal high in monoamine content had disappeared by 14 hours after meal consumption. Therefore, an extended overnight fast would eliminate dietary effects from samples drawn late in the morning hours. On the other hand, the use of a special diet in place of fasting is not ideal. Both this study and a previous investigation (12) show that an unpredictable effect of the presumably low monoamine meal is an elevation in pHVA in a third of the subjects tested. In conclusion, pHVA sampling should be preceded by a 14 hour fast.

2b) **Effects of physical activity on pHVA concentrations in human subjects**

Strenuous to moderate physical activity, such as exercising on a bicycle ergometer, induces a 15% elevation of pHVA that persists for one hour (12). This would suggest that such physical activity be restricted several hours before pHVA sampling. Six medically and psychiatrically healthy volunteers participated in a study examining the effects of mild physical activity on pHVA concentrations. Subjects changed position from recumbency, to standing up for fifteen minutes and later walked at a fixed pace for thirty minutes. The values of pHVA obtained during this "physical activity study day" were compared to pHVA values obtained at the same time points on a control day during which subjects maintained complete bed rest.

No significant difference between the two days and no interaction of the activity condition with time emerged (11). Analysis of the time points immediately following physical activity also indicated no effect on pHVA levels. Thus, although strenuous exercise should be restricted several hours before pHVA sampling, complete bed rest is not necessary.

3a) Relationship between severity of schizophrenic symptoms and ᵢHVA concentrations in drug-free patients.

Although the dopaminergic hypothesis of schizophrenia was first advanced almost three decades ago (13), the precise role of this neurotransmitter in schizophrenia has remained elusive. Considerable data, however, would agree that dopaminergic activity is more related to modulation of schizophrenic symptoms than to their initiation. DA activity might influence the symptom severity, rather than the presence or absence of schizophrenia. Consequently, a study was designed to examine the relationship between symptom severity and pHVA concentrations in schizophrenics withdrawn from neuroleptic drugs.

Thirty-five physically healthy, male schizophrenic patients meeting Research Diagnostic Criteria (RDC positive) for schizophrenia participated in this study. Severity of schizophrenic symptoms was assessed by two raters using the BPRS and CGI. Elements of the BPRS felt to reflect positive symptoms, i.e. hallucinations, unusual thoughts, suspiciousness, and bizarre behavior, were summed to form a positive symptom severity score. Following 14 hours of fasting and restricted physical activity, blood samples were drawn between 0830 and 1000. pHVA concentrations correlated with severity of schizophrenic symptoms as reflected by the CGI score (r = .46, p .003) total BPRS scores (r = .39, p .01) and positive symptom BPRS scores (r = .40, p .01).

3b) Comparison of ᵢHVA concentrations between drug free schizophrenic patients and normal control subjects.

In spite of considerable variation of pHVA concentrations in both normal control subjects and schizophrenic patients. pHVA measurements may nevertheless be able to distingush between normal controls and particular subgroups of schizophrenic patients. Fourteen male schizophrenic patients mean age 34.3 + 8.3 (SD) years, RDC positive for schizophrenia, participated in this study. The control group consisted of fourteen healthy volunteers, mean age 33.6 + 13.7 (SD) years, admitted 24 hours before the study. Schizophrenic patients were hospitalized and free of oral neuroleptics for at least three weeks, and free of depot neuroleptics for at least three months prior to study. CGI and BPRS scores on the day of the study were 3.8 and 38.4, respectively. The mean time ill (chronological age minus age at onset of illness) was approximately 10 years. Only one of the 14 patients had a documented response to neuroleptic treatment in a prospective standardized protocol. pHVA was sampled at hourly intervals between 10 p.m. and 10 a.m. the next morning. Patients were hospitalized for at least 3 weeks prior to the study, whereas the normal controls, although adapted to the sleep study room and observed to have two good nights of sleep, were in the hospital for only 24 hours prior to the study. This difference in itself could have been a nonspecific confounding nuisance. In order to rule out the possibility that the group differences are effected by the time spent in the hospital, eight normal controls had a sleep study 24 hours after admission, and a second sleep study following 3 weeks of hospitalization. No significant differences in pHVA concentrations were observed between the first and the second sleep study (mean average 9.0 ng/ml + 2.2 SD and 9.2 ng/ml + 1.4 SD respectively).

Average mean pHVA concentrations in schizophrenic patients were significantly lower than in normal controls (t = 2.38, p .02). Furthermore, pHVA values were lower in patients compared to controls at every time point during the 12 hour study (repeated measures analysis of variance F = 7.01, p 0.13). An examination of the relationship between pHVA concentrations and CGI/BPRS scores in the group of schizophrenic patients revealed a significant positive correlation. These results indicated that pHVA concentrations for a 12 hour period including the night were lower in chronic schizophrenic patients than in normal control subjects. These data are consistent with another very recent study showing that the 24 hour urinary excretion of DA and its metabolites was reduced in chronic schizophrenic patients as compared to control subjects (14). This finding is an apparent deviation from the idea that in schizophrenic patients, cerebral dopaminergic neurotransmission is inevitably augmented throughout the brain. Within the schizophrenic group, however, the more symptomatic patients had higher pHVA concentrations than the less ill schizophrenics. Such a finding is in agreement with the results of other investigations (15), and is consistent with a role for DA in modulating the severity of schizophrenic symptomatology. Finally, it must be realized that these results derive from a group of essentially treatment resistant patients, and that these results cannot be generalized to all schizophrenics (16). In support of the idea that chronic schizophrenic patients nonresponsive to neuroleptic treatment have lower pHVA concentrations, it has been reported that patients with the highest pHVA concentrations were the most neuroleptic responsive.

Given the correlations between pHVA and severity, it is quite conceivable that a group of very disturbed schizophrenics would have pHVA substantially higher than the current study population and would not be so readily distinguishable from controls. Severe hypothetical formulations related to the postulated dopaminergic abnormality in schizophrenia could account for the lower pHVA values in these schizophrenic patients. Clinical and postmortem investigations have indicated that in schizophrenic patients DA receptor density is augmented (17,18). Increased receptor density could account for both increased post-synaptic DA effects and, via a feedback mechanism, reduced central and peripheral DA turnover and HVA formation.

An entirely different model is suggested by the intriguing finding that ablation of DA tracts in the frontal cortex of rodents is accompanied by decreased frontal DA activity, increased DA receptor density in the caudate-putamen and increased presynaptic dopamine activity in the nigrostriatum (19). According to this model, lesions of DA projections to the cortex that produce cortical hypodopaminergic yield additional abnormalities of subcortical dopaminergic tracts, specifically, subcortical hyperdopaminergia. Although quite speculative, it is possible that the two major findings of the present report, low plasma HVA in schizophrenics relative to normal individuals in the face of positive correlation between pHVA and severity of schizophrenic symptoms, can be related to the relative contributions of the frontocortical and the subcortical DA systems to pHVA. If the frontal cortex is normally a prominent source of pHVA, a likelihood supported by the relatively rapid DA turnover and the total amount of tissue in this area (20), a less active frontal cortex in schizophrenics as a whole would lead to lower group mean levels of pHVA compared to normals. However, within the schizophrenic group, if the degree of manifest psychopathology is positively correlated to striatal or limbic DA activity, the severity of symptoms could also directly correlate with pHVA. Both increased DA transmission in subcortical regions (17,18) and decreased metabolic activity in frontal regions (21,10,22,23) has been demonstrated in schizophrenic patients.

REFERENCES

1.   Kendler KS, Heninger GR, Roth RH (1981).   Brain contribution to haloperidol-induced increase in plasma homovanillic acid. Eur J Pharmacol, 71, 321-326

2.   Kendler KS, Heninger GR, Roth RH (1982).   Influence of dopamine agonists on plasma and brain levels of homovanillic acid. Life Sci, 30, 2063-2069

3.   Medina NA, Giachetti A, Shore PA (1969).   On the physiological disposition and possible mechanism of the antihypertensive action of debrisoquin. Biochem Pharmac, 18:891-901

4.   Riddle MA, Leckman JF, Cohen DJ, Anderson M, Ort SI, Caruso KA, Shaywitz BA (1986).   Assessment of central dopaminergic function using plasma-free homovanillic acid after debrisoquin administration. J Neural Transm, 67:31-43

5.   Maas JW, Contreras SA, Bowden CL, Weintraub JE (1985). Effects of debrisoquin on CSF and plasma HVA concentrations in man.   Life Sci, 36:2163-2170

6.   Davila R, Zumarrage M, Perea K, Andia I, Friedhoff AJ (1987). Elevation of plasma homovanillic acid level can be detected within four hours after initiation of haloperidol treatment. Arch Gen Pscyh 44:837-838

7.   Friedhoff AJ (1986). A dopamine dependent restitutive system for the maintenance of mental normalcy.   Annals of the New York Academy of Sciences, 463:47-52

8.   Wolkowitz OM, Breier A, Doran A, Kelose J, Lucas P, Paul SM, Pickar D (1986).   Aprazolam Augmentation of Neuroleptic Antipsychotic Effects. Presented   at   the   American   College   of   Neuropsychopharmacology, Washington, DC

9.   Bannon MJ, Roth RH (1983). Pharmacology of mesocortical dopamine neurons. Pharmacol Rev, 35:53-68

10.   Weinberger DR (1987). Implication for normal brain development for the pathogenesis of schizophrenia. Arch Gen Psych, 44:660-669

11.   Davidson M, Giordani A, Mohs RC, Aryan T and Davis KL (1987b). Control   of   extraneous   factors   affecting   plasma   homovanillic   acid concentrations. Psychiatr Res, 20:307-312

12.   Kendler KS, Mohs RC, Davis KL.   The effects of diet and physical activity on plasma homovanillic acid in normal human subjects. Psychiatr Res, 8:215-223

13.   Carlsson A, Lindquist M (1963).   Effect of chloropromazine or haloperidol on formation of 3-methoxytyramine and normetanephrine in mouse brain. Acta Pharmacol Toxicol, 20:140-144

14.   Karoum F, Karson CN, Bigelow LB, Lawson WB, Wyatt RJ (1987). Preliminary evidence of reduced combined output of dopamine and its metabolites in chronic schizophrenia. Arch Gen Psych, 44:604-607

15.   Pickar D, Labarca R, Doran A, Wolkowitz GM, Roy A, Breier A, Linnoila M, Paul SM (1986). Longitudinal measurement of plasma

homovanillic acid levels in schizophrenic patients. Arch Gen Psych 43:669–676

16.   Keefe RSE, Mohs RC, Losonczy MF, Davidson M, Silverman JM, Kendler KS, Horvath TB, Nora R, Davis KL (1987). Characteristics of very poor outcome schizophrenia. Am J Psych, 144:889–895

17.   Seeman P, Vepian C, Bergerson C, Rieder P, Jellinger R, Gabriel E, Reynolds GP, Tourtellotte WV (1984). Bimodal distribution of dopamine receptor density in brains of schizophrenics. Sci 225:728–731

18.   Wong DF, Wagner HN, Tune LE, Dannals RF, Pearlson GD, Links JM, Tamminga CA, Broussolle EP, Ravert HT, Wilson AA, Young TK, Malat J, Williams JA, O'Tauma LA, Snyder SH, Kuhar MJ, Gjedde A (1986). Positron emission tomography reveals elevated D_2 dopamine receptors in drug-naive schizophrenics. Sci, 234:1558–1563

19.   Pycock CJ, Kerwin RW, Carte CJ (1980). Effect of lesion of cortical dopamine terminals on subcortical dopamine in rats. Nat, 286:74–77

20.   Bannon MJ, Reinhard JF, Jr., Bunney EF, Roth RH (1982). Unique response to antipsychotic drugs is due to absence of terminal receptors in mesocortical dopamine neurons. Nat, 296:444–446

21.   Weinberger DR, Berman KF, Zec RF (1986). Physiological dysfunction of dorsolateral prefrontal cortex in schizophrenia: regional cerebral blood flow (rCBF) evidence. Arch Gen Psych, 43:114–125

22.   Berman KF, Zec RF, Weinberger DR (1986). Physiological dysfunction of dorsolateral prefrontal cortex in schizophrenia. Role of medication, attention and mental effort. Arch Gen Psych, 43:114–125

23.   Volkow ND, Brodie DJ, Wolf AP, Gomez-Mont P, Cancro R, Van Gelder P, Russell JAJ, Overall J (1986). Brain organization in schizophrenia. J Cerebral Blood Flow and Metabolism, 6:441–446

21
Glucose-6-phosphate dehydrogenase deficiency and psychoses

A. Bocchetta, M. Del Zompo and G.U. Corsini

INTRODUCTION

Glucose-6-phosphate-dehydrogenase (G6PD) catalyzes the first step in the hexose monophosphate (HMP) shunt. This pathway serves to supply the cell with NADPH, the most important function being probably the maintainance of glutathione (GSH) in the reduced state. G6PD deficiency is the most common disease-producing enzyme deficiency of human beings, affecting about one hundred million people throughout the world. G6PD is subject to different mutations, and more than two hundred variants have been reported. Some of these cause a deficiency of enzyme activity in erythrocytes, leading to hemolitic anemia, usually upon exposure to an offending drug or toxin. Hemolysis may also occur during infection, diabetic acidosis, or when there is no known inciting cause. Some persons, perhaps carrying an additional genetic defect, show a particular sensitivity to fava beans (favism), whether upon ingestion or exposure to pollen. Different genetic variants of G6PD may have effects in tissues other than the erythrocytes, including the CNS.

The gene for G6PD is located on the X chromosome, closely linked to the color blindness region. The deficiency state is, therefore, a sex-linked trait. Affected males (hemizygotes) inherit the abnormal gene from their mothers, and their erythrocyte enzymatic activity, at least in the Mediterranean variant, is zero, like that of homozygous deficient females. Heterozygous females usually show intermediate values, but the activity may vary from zero to normal values because of inactivation of one of the two X chromosomes. Those who happen to have a high proportion of deficient cells resemble the deficient homozygotes and hemizygotes.

The possible relationship between G6PD deficiency and psychoses was first suggested by the case reports of two G6PD-deficient black men, who developed a temporary psychosis during the administration of primaquine [1]. Subsequently, G6PD deficiency was surveyed

in different populations of chronic schizophrenics. Dern et al
[2] found a normal overall prevalence of G6PD deficiency in schizophrenic black subjects in the Chicago area, but a statistically
significant difference between paranoid (low prevalence) and catatonic (high prevalence) subtypes of schizophrenia was reported for
both sexes. The finding was not confirmed in a survey of male black
patients in Alabama [3], while conflicting results were obtained in
New York [4]. The main reason for these discrepancies were probably
the unreliability of the diagnostic classification.

Several years after the first report, Nasr [5] observed the
case of a young black woman who suffered two acute paranoid psychotic episodes with quick resolution and suggestive evidence of G6PD-
related hemolysis after the ingestion of fava beans. Nasr proposed a
correlation between the enzymatic deficiency and an organic delirium
with paranoid features as opposed to an acute paranoid schizophrenia, considering the presence of alternating disorientation, the
excellent recovery of the patient, the absence of any family history
of mental illness, and the absence of clear stressful life events
leading to her psychosis.

In a subsequent preliminary report, Nasr et al [6] surveyed
G6PD activity in 104 patients, hospitalized in Illinois for unipolar
depression, bipolar disorder or schizophrenia. They found a statis-
tically- not-significant trend for an increased incidence of G6PD
deficiency in bipolar patients and in black patients, particularly
females.

We have recently reported a preliminary survey of G6PD deficiency in 161 affectively ill outpatients in Sardinia [7]. A significantly higher prevalence of the enzymatic deficiency was
found in bipolar females, particularly those suffering from hypomanic episodes (bipolar 2 depression), compared to controls.

Now, we report an extension of the previous survey. Moreover,
we report the incidental observation of acute psychotic episodes
with manic symptoms in three Sardinian men, with concomitant G6PD-
related hemolysis and family history of recurrent mental illness.

MATERIALS AND METHODS

The subjects in this study were 294 outpatients of the Lithium
Clinic of the University of Cagliari. Diagnoses were made separately
by two psychiatrists according to the Research Diagnostic Criteria
(RDC) [8]. The controls (n=60) were University employees not otherwise screened. Both patients and controls were Sardinian. G6PD
activity in erythrocytes was determined by the method of Kornberg
and Horecker [9]. Normal values are 131 ± 13 mU/10^9 erythrocytes

10 . Split-sample and test-retest reliabilities were both high. G6PD determination and psychiatric diagnoses were always done blind to one another. Informed consent was obtained from all concerned after the nature of the procedure had been fully explained.

RESULTS

Table 1 shows the number of patients with G6PD deficiency by sex and diagnosis. The distribution of enzymatic activity in our sample was similar to that observed in the general population in Sardinia [11]. Deficient males invariably showed activity lower than 3% of normal values, the only exception being a control subject with 28 mU/10^9 erythrocytes. Enzymatic activity in females varied from zero to normal values. An enrichment was observed below 10% of normal values. We selected this as the deficient group, which would predominantly represent homozygotes.

Females suffering from bipolar depression with hypomania (bipolar 2) exhibited a significantly higher prevalence of G6PD deficiency compared to controls. A trend for an increased prevalence, even though statistically not significant, was observed in bipolar 1 females and in manic schizoaffective males.

In addition, we reanalyzed for significance the data for female subjects after selection of different percentage levels of enzymatic activity, in order to include G6PD-deficient heterozygotes. An even higher difference (chi-square = 6.17, p < 0.025) was observed between bipolar 2 (12/32 = 37.5%) and control females (3/34 = 8.8%) with

Table 1. G6PD deficiency in psychiatric outpatients and controls

Diagnosis	No. deficient / No. subjects (%)			
	Males		Females	
Unipolar depression	1/14	(7.1)	2/45	(4.4)
Bipolar 2 depression (with hypomania)	2/14	(14.3)	9/32*	(28.1)
Bipolar 1 depression (with mania)	4/22	(18.2)	5/42	(11.9)
Schizoaffective disorder (depressive)	0/9	(0.0)	0/34	(0.0)
Schizoaffective disorder (manic)	14/39	(35.9)	1/43	(2.3)
Controls	4/26	(15.4)	2/34	(5.9)
Total	25/124	(20.2)	19/230	(8.3)

* significantly different from controls (chi-square = 4.38, p < 0.05)

enzymatic activity lower than 40% of normal values. Below this percentage level, significant difference (chi-square = 4.08, p < 0.05)
was also observed when comparing controls with bipolar 1 and bipolar
2 females taken as a single group (21/74 = 28.4%).

CASE REPORTS

Case I

A 36-year-old man had bipolar depression with mood-incongruent psychotic features since the age of 19. He was well without medication
in the last 3 years, when he became insomniac, hyperactive, more
talkative than usual, and manifested prodigality and persecutory
delusion. He was hospitalized and routine laboratory tests revealed
generally normal results, except for the presence of hemolysis (mild
anemia; urobilinogenuria). He was given haloperidol and chlorproma-
zine, and his symptoms resolved within 10 days.

On discharge, he was referred to the Lithium Clinic, where he
admitted to having eaten fava beans some days before the onset of
his symptoms. He was G6PD deficient, and had been hospitalized for
favism at age 6.

All his available relatives were personally interviewed and
tested for G6PD activity. His mother had had several depressive episodes and one hypomanic episode during tricyclic antidepressants.
Her G6PD activity was zero. She had favism at age 40. The three
patient's brothers had had recurrent depressive episodes. They were
G6PD deficient. One of them had had hypomanic episodes during antidepressants, and was taking lithium carbonate. He had favism at age
12. One sister (age 38) was psychiatrically normal. The other had a
single manic episode at age 25. Both sisters had intermediate enzymatic activity. The father was psychiatrically and enzymatically
normal.

Case II

A 28-year-old man, who had predominantly had depressive episodes
with depersonalization and persecutory delusion since the age of 15,
and some episodes of hyperactivity and elated mood, was referred to
the Lithium Clinic.

On admission, he was euphoric, hyperactive, overtalkative, and
manifested inflated self-esteem. Routine laboratory tests revealed
the presence of hemolysis (indirect bilirubin = 1.3 mg%; urobilinogenuria). His G6PD activity was zero. He admitted to having eaten

fava beans some days earlier. Other members of his family (mother, one sister, a maternal aunt, a maternal uncle) had been treated for recurrent mental illness, but were not available for direct inter-view or enzymatic determination.

Case III

A G6PD-deficient 25-year-old man had been under long-term lithium carbonate since the age of 22. He had had several schizoaffective episodes, both manic and depressive. We retrospectively analyzed the record of his first admission to mental hospital, and found the presence of hemolysis (mild anemia; total bilirubin = 1.9 mg%; uro-bilinogenuria). On that occasion, he had manic symptoms accompanied by delusions of grandeur and persecution. He was treated with chlor-promazine and haloperidol. The record did not mention any exposure to fava beans or hemolysis-inducing drugs. The patient denied any evidence of hemolysis in the past. One sister of his had suffered two schizophreniform episodes, but was not available for direct interview or G6PD determination.

DISCUSSION

The results of our survey and our case reports suggest that a corre-lation between G6PD deficiency and psychoses is to be searched in populations and families of patients suffering from acute psychotic episodes with affective symptoms.

Previously, G6PD activity was extensively surveyed only in populations of chronic schizophrenics, and the results were con-flicting [2-4]. On the other hand, the cases reported so far con-cerned two men with a temporary (not otherwise described) psychosis [1], and a woman diagnosed as acute paranoid schizophrenic, who also manifested depressive mood [5].

Several factors, including genetic heterogeneity and population stratifications, hinder the establishment of association between biological variables and disease, especially if common like affec-tive illness. According to the present report, we propose that sub-jects with a discrete, genetically determined, enzymatic defect such as G6PD deficiency, may be particularly susceptible to affective disorders. The presence of a sex effect (G6PD-deficient males appear to undergo severer psychotic episodes than females) must be con-firmed by more extensive surveys.

Several hypothesis may be attempted to explain an association between G6PD deficiency and psychoses. The first is that the enzy-

matic defect in erythrocytes, leading to anemia or hyperbilirubin-
emia, might play a causal role in the mental illness. It is known
that hemolysis can be toxic to the CNS, especially in the presence
of factors which enhance the passage of bilirubin through the blood-
brain barrier, like in kernicterus. The consequences may be different depending on severity and age of onset.

It is noteworthy that Meijer [12], examining for brain dysfunction 37 G6PD-deficient boys from one school and 37 paired matched
controls from the same school in Israel, found significantly more
hyperkynesis, writing problems, speech defects, aggression, and a
strong trend toward mood changes in the G6PD-deficient boys.

An alternative hypothesis is that the effects on erythrocytes
may be not related with those on the CNS. In the acute cases, hemolysis may have simply revealed exposure to oxidative stress.

In the favism-related cases, attention must be drawn to the
high L-dopa content of fava beans. The possible hemolytic effect
of L-dopa has not been confirmed [13] but it is known that L-dopa
can induce psychiatric symptoms including delirium, depression,
paranoid delusions, hallucinations, and hypomania during the treatment of parkinsonian patients [14]. It appears that a subgroup of
patients are at risk to develop severe affective reactions, including suicide. These patients are characterized by a previous history
of affective episodes and by a familial predisposition to affective
illness [15]. In this case G6PD deficiency may be not involved,
but L-dopa-induced potentiation of dopaminergic transmission may
suggest the possible link.

It is known that monoamine transmitters including dopamine can
stimulate the activity of the HMP shunt in brain, perhaps in relation with detoxication of reactive intermediates and free radicals
generated in the metabolism of the neurotransmitter substances [16].
Monoamine oxidase (MAO) activation appears to be the main mechanism
implicated in such a stimulation, via the production of aldehyde
derivatives of biogenic monoamines which are reduced to the corresponding alcohols by an NADPH-linked aldehyde reductase [17]. Moreover, the MAO-dependent oxidative deamination of dopamine generates
hydrogen peroxide, which is detoxicated via glutathione peroxidase
and NADPH-dependent glutathione reductase [18]. NADP levels are the
limiting factor of the activity of the HMP shunt.

G6PD deficiency is a condition where reduced glutathione and
NADPH are in poor supply, and detoxication of products of monoamine
metabolism may be defective, especially in the presence of enhanced
monoaminergic transmission or other oxidative stress. Consequently,
protein sulfhydryl groups cannot be maintained and CNS damage may
follow, whether due to loss of structural integrity of the lipid-
protein membrane, like in the erythrocyte, or to inhibition of key

enzyme-related functions.

Different mechanisms must be considered in the favism-related cases. Certain fractions of fava beans besides L-dopa have been implicated in the mechanism of hemolysis. For example, divicine is a possible candidate for its capability to generate hydrogen peroxide or free radicals.

As to the primaquine-related cases, it is known that anti-malarials can have CNS side effects, besides their peculiar propensity to induce hemolysis. Various psychoses have been particularly reported during administration of quinacrine [19]. Quinacrine can stimulate the HMP shunt and the related glutathione reduction in brain preparations [20].

Whatever the mechanism involved, the question rises whether G6PD-deficient subjects are particularly susceptible to affective episodes even in the absence of previous or concurrent hemolysis or exposure to identified specific environmental oxidants (drugs or fava beans). The issue is not academic, considering that about one hundred million people throughout the world are G6PD deficient.

Further studies are warranted in order to clarify other issues, such as the role of different G6PD variants and additional components, whether genetic or environmental. Accurate analysis of families of doubly affected individuals will provide interesting information.

We can attempt some speculation from published X-chromosome linkage studies. Risch and Baron [21], reanalyzing linkage data, have confirmed that close linkage of bipolar illness to the color-blindness-G6PD-region of X-chromosome appears to be present in some pedigrees. Linkage and association can be confounded when a disease-marker association is present. This may occur either if a disease-susceptibility locus is closely linked to the marker locus and in linkage disequilibrium [22] or if the associated marker allele is directly involved in the etiology of the disease, perhaps through epistatic interaction with another locus [23] . In the latter case, a submodel of the linkage model can be used, where the recombination fraction is specified as $\theta = 0.0$ and association between marker and disease susceptibility allele is complete [24].

Risch and Baron [21] have proposed linkage as opposed to association between affective illness and color blindness since the two traits have been found to be both in repulsion or in coupling in pedigrees with positive lod scores. The estimated recombination fraction between bipolar illness and color blindness is $\theta = 0.05$.

On the contrary, association cannot be excluded between G6PD deficiency and affective illness in the three informative families reported so far. Mendlewicz et al [25] reported a five-generation pedigree ascertained through a female patient of Persian Sephardic origin, whose son had been known to suffer from hemolytic anemia for many years. This is the single pedigree with the largest reported maximum lod score (5.11 at θ = 0.0).

We have recently reported a Sardinian pedigree informative for linkage between bipolar illness and G6PD deficiency [26]. The maximum lod score was 0.97 at θ = 0.0.

Baron et al [27] have reported a maximum lod score of 2.94 at θ = 0.0, in a Jewish pedigree originated from Iraq.

All affectively ill subjects examined in the three studies (total number = 23) invariably carried the G6PD-deficient allele.

We propose that G6PD variants should be searched in affectively ill individuals, when X-linked transmission is suspected, that is in one third of the bipolar population as estimated by Risch et al [28]. Moreover, defects in detoxication-related pathways should be considered as possible genetic components in affective disorders, even when the pattern of inheritance is not known.

REFERENCES

1.   Dern, RJ, Glynn, MF and Brewer, GJ (1962). Studies on the influence of hereditary G6PD deficiency in the expression of schizophrenic patterns. Clin Res, 10, 80

2.   Dern, RJ, Glynn, MF and Brewer, GJ (1963). Studies on the correlation of the genetically determined trait, glucose-6-phosphate dehydrogenase deficiency, with behavioral manifestations in schizophrenia. J Lab Clin Med, 62, 319

3.   Bowman, JE, Brewer, GJ, Frischer, H, Carter, JL, Eisenstein, RB and Bayrakci, C (1965). A re-evaluation of the relationship between glucose-6-phosphate dehydrogenase deficiency and the behavioral manifestations of schizophrenia. J Lab Clin Med, 65, 222

4.   Fieve, RR, Brauninger, G, Fleiss, J and Cohen, G (1965). Glucose-6-phosphate dehydrogenase deficiency and schizophrenic behavior. J Psychiat Res, 3, 255

5.   Nasr, SJ (1976). Glucose-6-phosphate dehydrogenase deficiency with psychosis. Arch Gen Psychiat, 33, 1202

6.   Nasr, SJ, Altman, E, Pscheidt, G and Meltzer, HY (1982). Glucose-6-phosphate dehydrogenase deficiency in a psychiatric population: a preliminary study. Biol Psychiatr, 17, 925

7.   Bocchetta, A, Del Zompo M, Martis, G and Corsini, GU (1985). Glucose-6-phosphate dehydrogenase deficiency in a Sardinian psychiatric population. IVth World Congress of Biological Psychiatry, abs. 231.1, p.183

8.   Spitzer, RL, Endicott, J and Robins, E (1977). Research Diagnostic Criteria, 3rd edition. (New York: Biometric Research, New York State Psychiatric Institute)

9. Kornberg, A and Horecker, BL (1955). Glucose-6-phosphate dehydrogenase. In: Colowick, SP and Kaplan, NO (eds.) "Methods in Enzymology". p.323. (New York: Academic Press)

10. Löhr, GW and Waller, HD (1974). Glucose-6-phosphate dehydrogenase. In: Bergmeyer, HU (ed.) "Methods of Enzymatic Analysis". p.636. (New York: Academic Press)

11. Siniscalco, M, Bernini, L, Filippi, G, Latte, B, Meera Khan, P, Piomelli, S and Rattazzi, M (1966). Population genetics of haemoglobin variants, thalassaemia and glucose-6-phosphate dehydrogenase deficiency, with particular reference to the malaria hypothesis. Bull WHO, 34, 379

12. Meijer, A (1984). Psychiatric problems of children with glucose-6-phosphate dehydrogenase deficiency. Int J Psychiatr Med, 14, 207

13. Gaetani, G, Salvidio, E, Pannacciulli, I, Ajmar, F and Paravidino, G (1970). Absence of haemolytic effects of L-Dopa on transfused G6PD-deficient erythrocytes. Experientia, 26, 785

14. Goodwin, FK, Murphy, DL, Brodie, HKH and Bunney, WE (1971). Levodopa: alterations in behavior. Clin Pharmacol Ther, 12, 383

15. Mendlewicz, J, Yahr, F and Yahr, MD (1976). Psychiatric disorders in Parkinson's disease treated with L-dopa: a genetic study. In: Birkmayer, W and Hornykiewicz, O (eds.) "Advances in Parkinsonism". p.103. (Basle: Editiones "Roche")

16. Hothersall, JS, Greenbaum, AL and McLean, P (1982). The functional significance of the pentose phosphate pathway in synaptosomes: protection against peroxidative damage by catecholamines and oxidants. J Neurochem, 39, 1325

17. Tabakoff, B, Groskopf, W, Anderson, R and Alivisatos, SGA (1974). Biogenic aldehyde metabolism, relation to pentose shunt activity in brain. Biochem Pharmacol, 23, 1707

18. Maker, HS, Weiss, C, Silides, DJ and Cohen, G (1981). Coupling of dopamine oxidation (monoamine oxidase activity) to glutathione oxidation via the generation of hydrogen peroxide in rat brain homogenates. J Neurochem, 36, 589

19. Engel, GL, Romano, J and Ferris, EB (1947). Effect of quinacrine (atabrine) on the central nervous system. Clinical and electroencephalografic studies. Arch Neurol Psychiat, 58, 337

20. Tarantino, LM and Hotta, SS (1974). Quinacrine stimulation of glutathione reduction dependent on the presence of a particulate brain subfraction. Proc Soc Exp Biol Med, 147, 887

21. Risch, N and Baron, M (1982). X-linkage and genetic heterogeneity in bipolar-related major affective illness: reanalysis of linkage data. Ann Hum Genet, 46, 153

22. Thomson, G and Bodmer, W (1977). The genetic analysis of HLA and disease associations. In: Dausset, J and Svejgaard, A (eds.) "HLA and Disease". p.84. (Copenhagen: Munksgaard)

23. Hodge, SE and Spence, MA (1981). Some epistatic two-locus models of disease. II. The confounding of linkage and association. Am J Hum Gen, 33, 396

24. Risch, N (1983). A general model for disease-marker association. Ann Hum Gen, 47, 245

25. Mendlewicz, J, Linkowski, P and Willmotte, J (1980). Linkage between glucose-6-phosphate dehydrogenase deficiency and manic-depressive psychosis. Br J Psychiat, 137, 337

26. Del Zompo, M, Bocchetta, A, Goldin, LR and Corsini, GU (1984). Linkage between X-chromosome markers and manic-depressive ilness. Acta Psychiatr Scand, 70, 282

27. Baron, M, Risch, N, Hamburger, R, Mandel, B, Kushner, S, Newman, M, Drumer, D and Belmaker, RH (1987). Genetic linkage between X-chromosome markers and bipolar affective illness. Nature, 326, 289

28. Risch, N, Baron, M and Mendlewicz, J (1986). Assessing the role of X-linked inheritance in bipolar-related major affective disorder. J Psychiatr Res, 20, 275

22
Early separation events and noradrenergic status during major depression

G. Conte, A. Calzeroni, A. Pennati, A. Terzi, A. Vita and E. Sacchetti

INTRODUCTION

The hypothesis that loss or separation from parents in childhood might constitute a vulnerability factor in the development of adult depressive episodes was suggested by Abraham in the first decades of this century (1).
In these latter decades this hypothesis has been tested and elaborated in an extraordinary number of studies, all of which attempt to demonstrate its validity.
Despite the different and rather broad criteria used for diagnosing depression and including patients in studies, most of studies have found that depressed patients have a higher incidence of loss or separation from parents in childhood than healthy controls, patients with other psychiatric disorders, and than patients with non-psychiatric diseases (for review, see 2). We recently replicated this finding by comparing a large sample of major depressives with sex and age-matched healthy subjects and non-affective psychiatric patients in the first controlled study to employ DSM-III criteria (3).
The low degree of specificity of our own and other findings, however, implies that the occurrence of certain early life events does not necessarily imply the development of depressive episodes. Moreover, the percentage of depressed patients with a positive history of early separation was not sufficiently high to support the hypothesis of a cause-effect relationship. Thus, childhood separation events are neither a necessary nor sufficient condition for the development of depressive episodes in adulthood; depressive disorders might develop despite the absence of early separation events. Consequently, it was suggested that affective disorders might be heterogeneous with respect to the presence of childhood separation. Several studies attempted to identify factors to explain this heterogeneity, that is the lack of an unfailing correspondence between early separation and the development of depression.

In the sixties and seventies, the severity of episodes, the presence of psychotic or neurotic features, the polarity of the disease, the sex of patients, personality profiles and proneness to suicidal behavior were considered as heterogeneity factors. But there were still no conclusive results despite the large-scale studies and design refinements (2). It should be noted, however, that there was no attempt to take the possible role of biological variables into account.
This is indeed striking since several primate studies did investigate the biological correlates of separation-induced depressive behavior. The results of these studies strongly indicate that the behavior of the mother-deprived infant is characterized by behavioral symptoms overlapping with those observed in children separated from their mothers (4). Consequently, this depressive-like behavior in monkeys has been considered as indicative of a depressive condition in experimental animals. In this depression model, the mother or peer-infant separation strongly suggests that the early separation event specifically involves both central and peripheral norepinephrine (NE) system activity (5). In addition, the pharmacological manipulation of the NE system was shown to strongly influence the behavioral response to separation events in experimental animals (5).
Taken together, these data indicate an association between early separation events and some changes in NE system activity during depression.
Further, since the early sixties, the most accreditated hypotheses on the pathophysiology of affective disorders point to the involvement of the noradrenergic system, either at the metabolic or pre or post-synaptic receptorial level, at least in a certain proportion of depressive patients (6).
The aim of the present study was to test the hypothesis that a childhood separation event in individuals who develop a major affective disorder in adulthood is associated with a distinct NE turnover pattern during depression as expressed by differences in urinary 3-methoxy-4-hydroxy-phenyl-ethylen-glyco (MHPG) excretion.

SUBJECTS AND METHODS

To be included in the study, patients had to meet the following criteria:
1) a diagnosis of major depression, bipolar or recurrent, according to DSM III criteria (7); all patients studied before the introduction of DSM-III met the Research Diagnostic Criteria (8) for Major Depressive Disorder (primary) and on the basis of the subsequent examination of clinical chart records of the index episode all of them also met the DSM III criteria;

2) a Hamilton Rating Scale for Depression score higher than 21 (9);

3) absence of somatic or neurological illness concomitant with the index episode; particular care was applied to exclude patients with diseases potentially able to influence the catecholaminergic turnover (for example, hypertension since this influences the relationship between arterial pressure and MHPG excretion in urine (10);

4) a drug-free period for at least four weeks before urine collection for MHPG measurement; only hypnotic benzodiazepines were allowed during this time;

5) strict compliance to a VMA, tyramine and methylxantine-free diet for five days, two before and three during urine collection.

Urine collection and gas-chromatographic determination of urinary MHPG have been described in detail elsewhere (11).

The MHPG value for each subject represents the mean value of three days of urine collection.

A further criterion for making patients eligible for the present study was the possibility of being able to document the occurrence or non-occurrence of the following life events before age 10:

1) death or departure of one parent or of the subject from the family for at least 6 months for any reason other than marital separation or divorce (work, war, boarding school, hospitalization);

2) severe disease, of one parent or of the subject, implying the risk of life or an excessively long period of bed rest with a resultant severe change of familial and/or social relationships.

The assessment of the occurrence or non-occurrence of life events was made during a euthymic period and was considered acceptable only when confirmed by a first degree relative or a close acquaintance of the subject with precise knowledge of his childhood.

Only the occurrence or non-occurrence of events was considered and no attempt was made to take the number and relative weight of single events into account.

Excluded from the study were all patients who had early life events clearly due to a psychiatric pathology in parents (such as repeated hospitalizations because of a psychiatric disorder or death by suicide), those with events suggestive of the existence of a possible underlying psychiatric disorder (such as separation from a parent because of divorce) also were excluded.

It was possible to document the occurrence or non-occurrence of early life events of 82 out of an original sample of 148 patients. These subjects met all the other criteria and had their urinary MHPG excretion measured.

Of the 82 patients, 33 were men and 49 women, 24 bipolar and 58 unipolar. The mean age $\pm$ SD was 47.5 ± 10.75 (range 23 to 71).

Since the urinary MHPG excretion for the total original sample was significantly higher in men than in women, the present study always used the sex variable as a correction factor for the various analyses performed.

RESULTS

Thirty-one (37.8%) of the 82 study patients reported the occurrence of at least one event of the type previously listed.
The mean age of patients with no early events was significantly higher than that of those who experienced such events: 49.47 ± 10.97 (SD) vs. 44.29 ± 9.7 (SD); Student t test: t = 2.164; p <.05. There was a nonsignificant trend for a higher incidence of patients with early life events among women than men: 22 (45%) out of 49 vs 9 (27%) out of 33; chi square = 2.6; p = n.s.
A positive history for negative early life events was similarly distributed among unipolar and bipolar patients: 21 (36%) of 58 unipolars vs 10 (42%) of 24 bipolars reported the occurrence of such events; chi square = .2; p = n.s..
Since the distribution curve of urinary MHPG levels in the original sample from which patients included in the present study came was skewed on the right both in men and in women, we utilized the log transformation of original MHPG values in our parametric analyses.
As already stated, the difference between men and women relative to urinary MHPG excretion compelled us to take the sex variable into account when the MHPG analyses were performed.
A two-way analysis of variance was undertaken on the urinary MHPG levels of our sample. Sex and the occurrence or non-occurrence of negative early life events were the two independent variables: both were highly significant (occurrence of early events: F = 14.8; d.f. 1,76; p <.0005; sex: F = 11.4; d.f. 1,76; p <.001). On the other hand, there was no interaction between these two variables (F = 2.683; d.f. 1,76; p = .102).

DISCUSSION

The incidence of patients with a positive history for childhood separation events in our sample was 37.8%. This figure overlaps with that of a collaborative study undertaken by the University of Florence on a larger sample of major depressives (34.4%)(3). This similarity may be taken as indicative of a lack of sampling bias in the present work.
The major result of our study was a highly significant

association between a low urinary MHPG excretion during
a major depressive episode and childhood separation
events.
We assumed that MHPG urinary excretion was indicative of
the individual's NE status. There is some evidence
confirming the validity of this assumption. First,
several recent reports have shown that the central and
peripheral NE systems are functionally related and work
in concert (12). Second, Linnoila et al. (13) found that
the urinary excretion of NE was highly correlated with
MHPG and other NE metabolites in a sample of major
depressives and healthy controls, thus indicating that
MHPG may be considered a reliable index of NE turnover.
In our study, we assumed the low urinary MHPG level as
indicative of a distinct pattern of the global NE system
turnover. Whatever the meaning of different NE turnover
patterns, we previously demonstrated that the pattern of
urinary MHPG excretion in major depressives is highly
reliable in independent depressive episodes, therefore
showing that this is an individual characteristic, at
least during depression (14). This finding represented a
sine qua non condition for undertaking the present
study. In a recurrent disease such as major depression a
lack of inter-episode reliability of the biological
variable under consideration would make nonsense of its
association with an antecedent environmental variable
such as childhood separation.
The finding that childhood separation events are
associated with a specific pattern of NE turnover is
consistent with results from experimental animal
studies. Kraemer et al. (5) showed that the intensity of
the behavioral response to social separation in infant
Rhesus monkeys was inversely correlated with baseline
CSF NE levels. Moreover, they showed that drug-induced
changes in the behavioral response to separation and in
CSF NE levels also were inversely correlated. On the
basis of these data these investigators hypothesized
that the individual's NE status and the pattern of his
behavioral response to separation is a trait
characteristic for that individual.
Suomi et al. (15) showed that the behavioral
disturbances following mother separation in 1-month old
rhesus monkeys were positively correlated with changes
in heart rate, taken as an index of autonomic system
reactivity. They further showed a positive correlation
between the 1-month heart rate change and evidence of
anxious behavior in novel situations when these
individuals became adults. These results were
interpreted as indicating both a behavioral reactivity
to environmental situations and an autonomic system
reactivity as peculiar traits for each individual.
Finally, Kraemer et al.(16) found an enhanced behavioral
responsiveness to pharmacological manipulations in adult
monkeys (for example, the administration of amphetamine)

if they had undergone social separation in childhood. The authors concluded that their data were suggestive of the long-standing impact of early social deprivation on brain neurochemical systems. This is not apparent under baseline conditions but shows up under particular conditions.

We are thus suggesting that an environmental childhood event, such as separation from mothers or peers, might act as a vulnerability factor for later destabilization of the NE system.

Our findings raise several questions and methodological problems. We chose the age 10 as the threshold level for considering the occurrence of separation events. This was decided not only because experimental animal studies have shown that the response to separation is age-dependent but also because it has been reported that central nervous system maturation is not complete until that age. It is obvious that an immature nervous and psychic substrate is more vulnerable and more prone to develop plastic changes than a mature, fully developed one. Last, selecting an age over 10 would have led to an overlap with the left tail of the distribution curve of age of onset of Major Affective Disorders.

It also should be mentioned, however, that several previous studies selected ages between 5 and 18 as the threshold for considering life events (2). Future studies certainly will consider the CNS maturation age more precisely and investigate the relationship between the biological variable and the occurrence of separation events at different ages so as to identify the maximum levels of specificity and sensibility.

Our study gave specific attention to including events with an objectively clear and datable separation significance. In so doing we were aware risking a low sensibility of our findings, but we preferred to guarantee a high specificity.

We are aware, however, we cannot give the same meaning and weight to all events included and that our approach does not take account of several peculiar separation events that may be meaningful for given subjects. What we tried to do was reach a compromise between an objective method considering a maximum number of events and the pitfalls of including events with subjective meanings that are impossible to check (high specificity/low sensibility vs high sensitivity/low specificity).

We did exclude all patients with a positive history of early events directly ensuing from a definite or probable psychopathologic disorder in their parents. This was done so as to avoid having a positive family history acting as a confounding variable relative to the association between negative early events and a specific NE turnover pattern. We even considered the possibility that in addition to causing stressful life events a positive family history for psychopathology might be associated with a reduced NE turnover trait, as some investigators have suggested (Conte G. and Sacchetti E., unpublished data): having included these subjects would have biased the resulting association because of the impossibility of separating the relative weights of life events and genetic vulnerability in the association with a low urinary MHPG output pattern.

To the best of our knowledge, this is the first study in which the relationship between an environmental and a biological variable has been investigated in humans. Since the sample investigated was rather large, the results obtained do not seem to be chance findings. In addition, our results are supported by data from experimental animal studies in a consistent way.

Our findings show that a particular environmental experience in childhood may be associated with a specific neurobiological pattern in later life, i.e., during a depressive state. This, too, is consistent with data from experimental animal studies. But the results, lend themselves to some interrelated hypotheses for future testing.

First, early separation events might be viewed as acting as sensitization factors that lead to the development of depressive disorders associated with a peculiar noradrenergic function pattern in a certain proportion of individuals.

In the last decade, in fact, a kindling phenomenon has been hypothesized to underlie the development of affective disorders (17).

Nonetheless, while this hypothesis is rather stimulating and supported by robust experimental results no kindling factor has yet been reported as predisposing factor for major affective disorder in humans. Our study could be considered a good preliminary in this research area.

Second, different research groups have suggested that major affective disorder is biologically heterogeneous. Different patterns of NE system activity have been found to characterize various subgroups of patients with major affective disorders showing distinct clinical, familial and neuromorphologic features (18).

The finding that the occurrence of early separation events is more frequently associated with a distinct pattern of NE turnover during depression might represent a further feature in characterizing distinct subgroups of affective patients. The demonstration of a relationship between early life events and a reduced NE turnover in major depression is the first attempt to validate different biological subgroups of major depressives according to parameters outside of the intrinsic function of the NE system. In previous studies on this problem, the activity of the NE system has been always focused on variables more or less directly related to it such as the response to antidepressants with a greater or lesser selective action on the NE system, the response of some hormones under more or less clear monoaminergic control, and the function of adrenergic receptors.

A third hypothesis is that a specific link associates separation events in childhood and an alteration of the central NE system, independent of the individual's diagnosis. In this regard it should be noted that the incidence of early separation events has repeatdly been shown to be significantly higher in patients with panic attack disorder than in healthy controls (19). Indeed, the central NE system has been claimed to play a significant role in the pathophysiology of this as well as affective disorder (20). This would imply that the relationship between early life events and a peculiar NE status is not specific to a certain diagnosis but must be looked for in other disorders.

REFERENCES

1) Abraham, K (1966). A short study of the development of the libido viewed in the light of mental disorders. In: "On character and libido development". p.67 (New York: W.W. Norton & Co.)

2) Lloyd, C (1980). Life events and depressive disorder reviewed. 1 Events as predisposing factors. Arch Gen Psychiatry, 37, 529

3) Faravelli, C, Sacchetti, E, Ambonetti, A et al (1986). Early life events and affective disorders revisited. Br J Psychiatry, 148, 288

4) Coe, CL and Levine, S (1981). Normal responses to mother-infant separation in nonhuman primates. In: Klein, DE and Rabkin, JG (eds.) "Anxiety. New research and changing concepts". p.155 (New York: Raven Press)

5) Kraemer, GW, Ebert, MH, Lake, R et al (1984). Cerebrospinal fluid measures of neurotransmitter changes associated with pharmacological alteration of the despair response to social separation in Rhesus monkeys. Psychiatry Res, 11/4, 303

6) Siever LJ (1987). Role of noradrenergic mechanisms in the etiology of the affective disorders. In: Meltzer, NY (ed.) "Psychopharmacology: the third generation of progress". p.493. (New York: Raven Press)

7) American Psychiatric Association Committee on Nomenclature and Statistics (1980). "Diagnostic and Statistical Manual of Mental Disorders, 3rd ed.". (Washington, DC: APA)

8) Feighner, JP, Robins, E, Guze, SB et al (1972) Diagnostic criteria for use in psychiatric research. Arch Gen Psychiatry, 26, 57

9) Hamilton, M (1960). A Rating Scale for Depression. J Neurol Neurosurg Psychiat, 23, 56

10) Potter, WZ, Muscettola, G and Goodwin, FK (1983).

Sources of variance in clinical studies of MHPG. In: Maas, JW (ed.) "MHPG: basic mechanisms and psychopathology". p.145. (Orlando: Academic Press)

11) Sacchetti, E, Conte, G, Pennati, A et al (1985). Platelet alpha-2-adrenoceptors in Major Depression: relationship with urinary MHPG and age of onset. J Psychiat Res, 19, 579

12) Leckman, F and Maas, JW (1983). Preliminary characterization of plasma MHPG in man. In: Maas, JW (ed.) "MHPG: basic mechanisms and psychopathology". p.107. (Orlando: Academic Press)

13) Linnoila, M, Karoun, F and Potter, WZ (1982). High correlation of norepinephrine and its major metabolite excretion rates. Arch Gen Psychiatry, 39, 521

14) Conte, G, Vita, A and Sacchetti, E (1988). Inter-episode reliability of urinary MHPG excretion in Major Depression. Biol Psychiatry, 24, 240

15) Suomi, SJ, Kraemer, GW, Baysinger, CM et al (1981). Inherited and experimental factors associated with individual differences in anxious behavior displayed by Rhesus monkeys. In: Klein, DF and Rabkin JG (eds.) "Anxiety. New research and changing concepts". p.155. (New York: Raven Press)

16) Kraemer, GW, Ebert, MH, Lake, CR et al (1984). Hypersensitivity to d-amphetamine several years after early social deprivation in Rhesus monkeys. Psychopharmacology, 82, 266

17) Post, RM, Rubinow, DR and Bellenger, JC (1984). Conditioning, sensitization and kindling: implication for the course of affective illness. In: Post, RM and Bollenger, JC (eds.) "Neurobiology of mood disorders". p.432. (Baltimore: Williams & Wilkins)

18) Vita, A, Sacchetti, E, Conte, G et al (1985). Heterogeneity of Major Affective Disorders. Biological and clinical evidences. L'Encéphale, 11, 71

19) Klein, DF (1981). Anxiety reconceptualized. In: Klein, DF and Rabkin JG (eds.) "Anxiety. New research and changing concepts". p.235. (New York: Raven Press)

20) Redmond, DE Jr. (1987). Studies of the nucleus coeruleus in monkeys and hypotheses for neuropsychopharmacology. In: Meltzer, NY (ed.) "Psycho-pharmacology: the third generation of progress" p.493. (New York: Raven Press)

Supported by CNR contract n. 87 00438.56 and Regione Lombardia grant n. 901.

23
Vulnerability and plasticity of monoamine neurotransmitter systems in affective and personality disorders

L.J. Siever, E.F. Coccaro, M. Davidson, L. Howard, L. Harter and K.L. Davis

INTRODUCTION

The monoamine neurotransmitter systems have attracted investigative interest in the last two decades for their putative role in the pathophysiology of two major DSM-III Axis I syndromes: the affective and schizophrenic disorders. The "catecholamine hypothesis" and the "dopamine hypothesis" of schizophrenia were based in part on some of the known pharmacologic effects of the antidepressants, e.g., catecholamine re-uptake antagonism, and the neuroleptics, e.g., dopamine receptor antagonism. Evidence supporting these hypotheses has not always been consistent, but has been sufficient to suggest some disturbances of the monoamine systems are associated with these DSM-III Axis I diagnosis, e.g., abnormalities of noradrenergic metabolite concentrations in the affective disorders and of dopaminergic metabolite concentrations in schizophrenia. More recent data suggests a "serotonergic hypothesis" of impulsivity/aggression, e.g., decreased serotonin metabolites in cerebrospinal fluid (CSF) of aggressive and suicide-attempting patients.

Function and role of monoamine systems: Can they be linked to dimensions of psychopathology?

While advances in our knowledge of the biology of the major psychiatric disorders have pointed to neurotransmitter disturbances in these disorders, a sharper understanding of the physiologic function of these monoamine systems has been emerging from neurobiologic studies. An increased comprehension of the functional role of the monoamine systems may point the way to delineating symptomatic dimensions associated with their dysfunction. The noradrenergic system, for example, seems to be phasic arousal system that increases cortical information processing in response to unexpected

231

or novel stimuli, particularly those that are potential threats to the organism [1,2]. The locus ceruleus, the major noradrenergic cell nucleus, has a robust firing rate during active goal-directed behavior but is quiescent during the vegetative functions of feeding, grooming, and sleeping [3]. Pre-clinical studies suggest that abnormalities in noradrenergic function accompany substantial negative environmental events over which the organism has no control, e.g., uncontrollable shock or prolonged separations [4]. In an animal model of uncontrollable shock, animals show disturbances in sleep, feeding, and activity paralleling many of the symptoms of major depressive disorder [4]. Thus, it is plausible to hypothesize that noradrenergic dysfunction in the affective disorders may be associated with the disturbances in a dimension of arousal and modulation encompassing both active goal-directed behavior and vegetative functions.

The dopamine system seems to be involved in the control of exploratory behavior in a novel environment while excessive activity may result in exaggerated, stereotypic behavior in human and animal studies [5]. In humans, modulation of frontal cortical activity by ascending mesocortical dopaminergic tracts may be important in screening, selecting, and processing incoming information for the purpose of formulating appropriate response strategies. Disturbances in dopaminergic function may thus be associated with a dimension of disordered attention/information-processing perhaps resulting in faulty reality-testing or even psychosis, as observed in schizophrenia [6,7].

The serotonergic system seems to be involved in the inhibition of aggressive behavior and the suppression/extinction of behavior with aversive consequences. Lesions of serotonergic neurons in animal studies, result in disinhibited muricidal behavior in rats and failure to learn to suppress punished behavior [8,9]. Disturbances in serotonergic activity might then be hypothesized to be associated with a dimension of excessive impulsive aggressive behavior with failures in the suppression of behaviors that lead to punishment or in "learning" not to engage in such behaviors, as observed in impulsive/sociopathic disorders. While these descriptions of the role of these neurotransmitter systems and possible symptomatic concomitants of their dysfunction are necessarily simplistic at this time, they at least suggest that disturbances in individual neurotransmitter systems have partially specific behavioral correlates.

Predisposition or vulnerability to psychopathology

These considerations thus raise the possibility that differences in specific neurotransmitter system activity might confer a predisposition to specific dimensions of psychopathology. These vulnerabilities may be severe in the case of chronic schizophrenia, bipolar affective disorders, and impulse disorders, but may be observed in attenuated forms in the personality disorders, e.g., milder defects in reality testing in schizotypal personality disorder, mood lability in borderline personality disorder, and impulsive behavior in borderline personality disorder.

Familial and adoptive studies suggest that there may be a genetic basis to the individual differences in specific dimensions of psychopathology. For example, the vulnerability to schizophrenia and related disorders, including schizotypal personality, seems to be genetically transmitted [10]. A genetic basis has been established for affective disorders that may extend to include some forms of borderline personality disorders [11]. Adoptive studies suggest a genetic basis for sociopathic and hysterical behaviors that might be encompassed in a dimension of impulsivity [12].

The manifestation of the vulnerability will depend both on environmental circumstances and the state of maturation of the individual, and thus may appear only at specific junctures in the development of the genetically predisposed. Affective disorders appear episodically with relative remissions between episodes, often in midlife or later, while schizophrenic and impulse disorders usually occur in young adulthood with a more tonic course, but may also be associated with periodic exacerbations. Any pathophysiologic mechanisms for these disorders and their genetic antecedents must accommodate their relatively phasic nature and partial environmental responsiveness.

Psychiatric disorders as dysregulation disorders

One model that accommodates these characteristics of psychiatric disorders is a dysregulation model. A dysregulation hypothesis of a psychiatric disorder posits that a failure in the homeostatic regulation of a neurotransmitter system's activity rather than a simple over or under-activity of that system is associated with the pathophysiology of these disorders. In psychiatric illnesses such as recurrent affective disorders, patients may be entirely well between episodes. The trait vulnerability to the symptomatic episode, i.e., an affective episode, must be present in these patients in the remitted state. However, other "state-dependent" biologic abnormalities must be associated with the symptomatic state, i.e., the affective episode. It might then be expected that a reduced efficiency at a regulatory state may be "silent" under non-stressful circumstances, but, with provocation by, for example, increased demand or stress on the system, the relevant neurotransmitter neurohormonal systems may become dysregulated, i.e., erratic, poorly modulated, and environmentally inappropriate in their activity. Efficacious pharmacologic treatments may restabilize and thus improve the efficiency of these systems.

This model calls for an investigation of state independent or trait correlates of the vulnerability to the specific psychiatric disorder that reflect an underlying regulatory defect. Such a correlate may ultimately aid in understanding the pathophysiology and genetic antecedents of the disorder. However, it is also important to identify the state-dependent configuration of neurotransmitter/neuromodulator alterations associated with the overt symptomatic episode, e.g., depression, to understand the final common of pathophysiology.

Such a model is quite consistent with the known pathophysiology a number of medical illness (Table 1). For example, an inappropriately reduced number of cholesterol receptors has been shown to be a genetically determined antecedent to hypercholesterolemia and ultimately to atherosclerosis in some individuals [13]. This genetic predisposition may be amplified or diminished by the amounts of cholesterol in the diet. Although the atherosclerosis in these individuals may be symptomatically "silent", its presence predisposes to an acute cardiovascular event such as myocardial infarction or stroke. However, the antecedent defect may be revealed in the high cholesterol levels that accompany this condition. Similarly, genetically determined defective insulin reception regulation may result in maturity-onset diabetes although the expression of the illness may depend on environmental factors such as diet and/or exercise [14]. However, even in the absence of the polyuria and polydipsia accompanying the acute decompensation of the diabetes, hyperglycemia will be detectable as a trait correlate of the illness. Analogously, then, defective neurotransmitter receptor regulation, perhaps on a genetic basis, might be apparent on provocative challenge in the non-depressed (remitted) state, when clinical symptoms may be silent.

Although the nature of such a neurotransmitter receptor defect remains unknown, it might be hypothesized that the plasticity or control of receptor sensitivity might be altered in some of the major psychiatric syndromes, e.g., major affective disorder. Thus, defects in receptor coupling and/or membrane fluidity may result in inappropriate up-and down-regulation of the relevant receptor systems. For example, relatively defective coupling of an adrenergic receptor recognition site to the G-protein mediating its effect on adenylate cyclase might result in uncoupled low affinity receptors with less capacity to modulate adenylate cyclase activity and decreased receptor recognition site phosphorylation and internalization, "locking" the receptor into a dysfunctional conformation. These considerations, although speculative, point to future directions for the investigation of neurotransmitter receptor regulation.

Specific examples of possible dysregulation of the different monoamine systems in psychiatric syndromes and their association with psychopathology will be presented below.

STUDIES OF MONOAMINE SYSTEMS IN PSYCHIATRIC SYNDROMES

The Noradrenergic System in Affective Disorder

Background. The noradrenergic system appears to be dysregulated in depression and meets a number of criteria for dysregulation of this system [15]. A blunted growth hormone response to clonidine, which appears to reflect post-synaptic hypothalamic receptor responsiveness, is suggested in studies of acute depressed patients and preliminary studies of remitted depressed patients suggesting the possibility that the blunted GH response to clonidine is a state-independent or trait correlate of depression [15,16].

As the alpha-2-adrenergic receptor plays a regulatory role in both modulating pre-synaptic noradrenergic output and hyperpolarizing other neuronal systems at post-synaptic receptor sites, defective alpha-2-adrenergic receptor function may represent a disturbance in inhibitory control mechanisms of noradrenergic release/metabolism as well as reduced efficiency of post-synaptic receptor effects to reduce "noise" in other neuronal systems. The net result would be a reduction in the potential effectiveness of norepinephrine's hypothesized function as an enabling system which enhances the "signal-to-noise" ratio of incoming stimuli. Such a reduced efficiency might become particularly important when the demand on the noradrenergic system is increased as, for example, by chronic stress and would result in the poorly modulated or dysregulated noradrenergic output observed in the acutely depressed state [15]. Thus, reduced responsiveness of the alpha-2-adrenergic receptor system might contribute to the vulnerability to depression.

For these reasons, alpha-2-adrenergic function was evaluated by measuring the growth hormone (GH) response to clonidine and pre-synaptic output of the noradrenergic system was assessed by hourly measures of MHPG throughout the day in both acute and remitted depressed patients.

Method. Male veteran patients with a current or past history of depression were clinically identified and diagnosed using the Schedule for Affective Disorders and Schizophrenia (SADS) [17] by two reliable raters

(K=0.93).All patients received a comprehensive medical evaluation and patients with systemic medical illness were excluded from the study. Alcohol and drug dependence were also grounds for exclusion from study. Acute depressed patients were required to prospectively meet Research Diagnostic Criteria (RDC) by direct clinical assessment for the duration of the two week medication-free and testing periods, while remitted patients must have been out of the hospital and in clinical remission for at least six months and scored less than 12 on the Hamilton Depression Rating Scale (HDRS) for three weeks prior to study.   All patients were medication-free for at least two weeks prior to testing, although the average interval off medication was much longer (54 ± 34 days), and were on a low monoamine diet for three days prior to and throughout the tests.

A baseline study of noradrenergic output indices was performed after the insertion of an intravenous line at 8 AM with hourly samples for plasma MHPG obtained between 10 AM and 6 PM. Clonidine was administered on a later day at a dose of 2mcg/kg intravenously at 10 AM after a two-hour acclimatization to the intravenous line and baseline blood samples were drawn. Samples for GH were obtained at 15 minute intervals for one hour following the infusion.

Results. The prevalence of blunted growth hormone responses (<5 ng/ml) to clonidine were almost identical between the acute depressed patients (n=28, 68%) and remitted depressed patients (n=15,67%) and both groups significantly differed from age-matched controls (n=12,25%) (Fisher's Exact Test,p<0.05). In contrast, the pattern of sequential plasma MHPG concentrations on the baseline study was erratic and apparently phase-shifted in the acute depressed patients but tended to normalize in remitted depressed patients. Under basal, resting conditions, the acutely depressed patients showed an earlier MHPG peak (13:08 +/- 2.43 hrs.) in comparison to normal controls (15:32 +/- 2.27 hrs.) with remitted patients (14:54 =/-1.36 hrs.) peaking close to the normal controls. Although mean plasma MHPG concentrations at 10:00 AM did not differ between groups, the variance between acutely depressed individuals was also greater than in the remitted patients (F=2.2,p<0.05).

Discussion. The blunted GH response to clonidine, a putative indicator of alpha-2-adrenergic receptor responsiveness, appears to represent a state-independent correlate of depression. In contrast, indices of noradrenergic activity such as plasma 3-methoxy-4-hydroxyphenylglycol (MHPG) show a pattern of increased erratic release in the acute state of depression, but tend to normalize in the remitted state.   Thus, noradrenergic output seems to be abnormal in a state-dependent fashion.

These results are consistent with the possibility that at least one defective control site, the alpha-2-adrenergic receptor, is altered in depressed patients compared to controls in a state-independent manner as suggested by criteria for a dysregulation hypothesis. Basal output of the system in the depressed state showed a loss of normal circadian rhythmicities with a phase advanced erratic rhythm, also consistent with the dysregulation model.

The dopaminergic system in schizophrenia-related disorders

Background. The dopamine system has been implicated as being associated with psychotic symptoms in the schizophrenia-related disorders on the basis of the correlation between the therapeutic efficacy of neuroleptic medications and their potency as dopamine antagonists [18] and between concentrations of plasma homovanillic acid (HVA) and severity of psychotic symptoms in schizophrenic patients [19,20]. The schizophrenia-related disorders extend beyond more, restrictive definitions of chronic schizophrenia to encompass the milder schizophrenia-related personality disorders. The DSM-III prototype for the schizophrenia-related personality disorders is schizotypal personality disorder, a disorder which preliminary evidence suggests has a phenomenological [21], genetic [22], biologic [23], prognostic [24], and treatment response [25,26] relationship to chronic schizophrenia.

Schizotypal personality disorder patients are less likely to have received long-term neuroleptic medication, to have been chronically hospitalized, or to have experienced the deteriorating effects of a severe, chronic psychotic illness. Thus, they provide the opportunity to study the relationship between the dopaminergic system activity and psychotic symptoms with less interference from these potentially confounding artifacts. While CSF concentrations of HVA have not generally distinguished schizophrenic from control populations, nor have the magnitude of HVA concentrations been consistently associated with specific symptomology [27], there have been suggestions of a positive association between CSF HVA concentrations and psychosis [28,29] and of a negative association between CSF HVA concentrations and "negative" symptoms of schizophrenia such as lassitude, decreased social interest, and poor prognosis [30]. More consistent relationships between these variables may have been obscured by the use of probenecid and by the artifacts inevitably associated with the study of schizophrenic patients.

Thus, it was of interest to study CSF HVA concentrations and their relationship to psychotic-like symptoms in schizotypal personality disorder patients and comparably ill controls, patients with other non-schizophrenia-related disorders.

Method. Male veteran patients clinically diagnosed as having a primary diagnosis of DSM-III personality disorder were diagnosed for Research Diagnostic Criteria (RDC) using the Schedule for Affective Disorders and Schizophrenia (SADS) (interrater reliability : K=0.93 for depression, K=0.80 for schizophrenia) and by the Schedule for Interviewing DSM-III Personality Disorders (SIDP) [31] with two raters independently but simultaneously rating the patient (interrater reliability for schizotypal personality disorder, K=0.77) and a third rater interviewing an informant close to the patient. A final diagnosis was determined by consensus. Test-retest reliability of the diagnosis of schizotypal personality disorder with an interval of at least six months was robust in a subsample of these patients (K=0.77).

Individual schizotypal personality disorder criteria were derived from the SIDP for all patients. Magical thinking, recurrent illusions, and ideas of reference were considered "positive" or psychotic-like symptoms. The Chapman psychosis-proneness scales including the physical anhedonia, social anhedonia, and perceptual aberration scales were administered to the patients to evaluate psychotic-like characteristics (32). The Minnesota Multiphasic Personality Inventory (MMPI) was also administered to evaluate schizophrenia -related psychopathology (paranoia, schizophrenia, and F scales) as well as other dimensions of psychopathology. The Speilberger State-Trait Anxiety Inventory (STAI) [33] and Buss-Durkee Hostility Inventory (BDHI) [34] were also administered to determine whether any relationship of CSF HVA concentrations to psychotic-like symptoms, if observed, might be attributable to less specific factors of anxiety or hostility.

Results. Schizotypal personality disorder patients had significantly increased concentrations of CSF HVA compared to patients with other personality disorders (p<0.01, Mann-Whitney U-test)(Figure 1). CSF HVA concentrations correlated with CSF 5-HIAA (r=0.64, df=11, p<0.05), but not with CSF MHPG (r=0.30, df=11, NS). There were no significant correlations between age, number of days off medication, duration of illness, or number of hospitalizations for the entire sample, although a negative correlation between CSF HVA concentrations and age was observed in the schizotypal subgroup (r = -0.81, df = 4, p < 0.05). The criteria reflecting psychotic-like symptoms were positively associated with CSF HVA concentrations [recurrent illusions (F = 13.6, df = (8,4), p < 0.05); ideas of reference (F = 27.1, df =

(8,4), p < 0.01); and magical thinking (F = 9.5, df = (8,4), p < 0.05)], while one "negative" symptom [inadequate rapport (F = 33.8, df = 8,4, p < 0.005)] was negatively related to CSF HVA concentrations by a multiple linear regressions analysis. CSF HVA concentrations correlated positively with scores on the Chapman physical anhedonia scale (rho = 0.77, df = 11, p < 0.005), the Chapman perceptual abberations scale (rho = 0.57, df = 11, p < 0.05), the MMPI F scale (rho = 0.69, df = 11, p < 0.05), the MMPI paranoia scale (rho = 0.65, df = 11, p < 0.02), and the MMPI schizophrenia scale (rho = 0.59, df = 11, p < 0.05). No significant correlations were observed for any of the other scales.

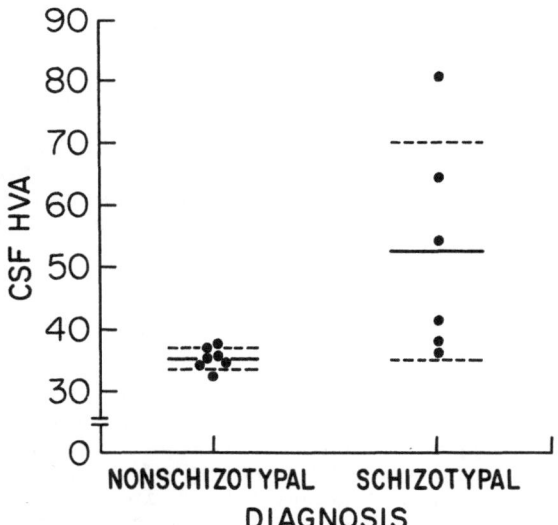

Fig. 1   CSF HVA concentrations (ng/ml) in schizotypal and other non-schizophrenia-related personality disorder patients.

Discussion. The results, although quite preliminary, raise the possibility that CSF HVA concentrations may be increased in at least a subgroup of schizotypal personality disorder patients and that this increase may be associated with the degree of psychotic-like symptoms. The results hint at the possibility that "negative" or defect symptoms, on the other hand, are not associated with increases and may be associate with decreases in CSF HVA. While studies of CSF HVA concentrations have not been entirely consistent in schizophrenic patients [27], possibly because of the confounding artifacts of probenecid administration, medication history, and duration of severe illness, several studies raise the possibility that CSF HVA concentrations may be positively associated with psychosis or paranoia [28,29] and negatively associated with negative symptoms [30]. Studies of plasma HVA provide stronger support for a relationship between dopaminergic activity and psychotic symptoms [19,20]. However, concentrations of HVA in plasma and urine [35,36] tend to be decreased in schizophrenic patients, while these results suggest increases in CSF HVA concentrations in schizotypal patients, but additionally hint there may be decreases in CSF HVA with increasing age in the schizotypal patients. Could reductions in dopaminergic metabolites in schizophrenic patients result from chronic, increased dopaminergic activity associated with prolonged schizophrenic psychosis? Young schizotypal patients who are not chronically psychotic might show increases in dopamine metabolites associated with psychotic-like symptoms. These considerations are speculative, but could be tested by longitudinal studies of both schizophrenic and schizotypal patients using comparable indices of dopaminergic activity.

The serotonergic system in impulsive, irritable aggressive personality

Background. Reduced measures of cerebrospinal fluid 5-hydroxynidoleacetic acid (CSF 5-HIAA) in patients with depression [37-40], schizophrenia [41,42], and personality disorder [43-46] and reduced binding of 3-{H}-imipramine in cortex [47] and hypothalamus [48] in suicide victims suggest that dysregulation of the central serotonergic system may be associated with behaviors related to physical aggression, impulsiveness, and suicide.

The actual nature of possible central serotonergic dysregulation in man is unclear. Data from available studies are generally interpreted as reflecting a relationship between diminished central serotonergic tone and behaviors related to physical aggression, impulsiveness, and suicide. However, while measures of putative pre-synaptic function (e.g. CSF 5-HIAA; cortical/hypothalamic 3-{H}-IMI binding) are reduced, measures of putative post-synaptic function (e.g. cortical 5-HT-2 binding; cortisol response to 5-HTP challenge), assessed in other patients, are elevated [49,50]. Thus, the "net" physiologic effect of diminished pre-synaptic serotonin output and increased post-synaptic receptor sensitivity is not known at the present time.

A promising index of overall central serotonergic function may be the prolactin response to acute challenge with the serotonergic releaser/uptake inhibitor fenfluramine. Fenfluramine activates the central serotonergic system through both pre- and post-synaptic mechanisms and may therefore reflect the sum total of more specific synaptic mechanisms (i.e. "net") central 5-HT function in the neurons involved in the hypothalamic-pituitary axis [51]. Thus, it was of interest to study prolactin responses to fenfluramine of patients with DSM-III personality disorder in order to assess the overall status of the central serotonergic system in patients with varying degrees of aggressiveness, impulsiveness, and suicidal behavior.

Methods. Twenty medically healthy male veteran patients clinically diagnosed as having a primary diagnosis of DSM-III personality disorder (see previous section on "Dopamine and schizophrenia-related disorders") gave informed consent to undergo a fenfluramine challenge. All studies were preceded by a drug-free period of two weeks, or more, and three-days of a low monoamine diet. After an overnight fast, the procedure began with the insertion of an indwelling intravenous catheter into a forearm vein at 8 am. Subjects then remained awake, supine and fasting until 3 pm. Samples for baseline plasma prolactin (PRL) were obtained at 9:45 and at 9:55 am (15 and 5 min.)> Fenfluramine 60 mg po was administered at 10 am (0 min.) and post-fenfluramine samples for plasma PRL were obtained hourly until 3 pm (+300 min.). Plasma PRL was assayed by RIA [52] and peak delta PRL (peak PRL minus the averaged baseline PRL) after fenfluramine was used as the measure of maximal PRL responsiveness to the serotonergic challenge; this measure correlated highly with the overall PRL response to fenfluramine (i.e. delta area under the curve: $r = 0.96$). Measures of aggression an impulsiveness were assessed with the Brown-Goodwin Assessment for History of Lifetime Aggression [44], The Buss-Durkee Hostility Inventory [53], The Psychopathic Deviance subscale of the MMPI [54], and the Barratt Impulsiveness Inventory [55]: history of suicide attempt was assessed during the SADS evaluation according to the criteria of Asberg et al. [37].

Results. Significant negative correlations were found between peak delta PRL responses to fenfluramine and Brown-Goodwin "Aggression" ($r = -0.57$, $p < 0.012$), Buss-Durkee "Motor Aggression" ($r = -0.52$, $p < 0.019$) and Barratt Total "Impulsiveness" ($r = -0.48$, $p < 0.03$). Significant negative correlations were also found for:

a) two subscales of Buss-Durkee "Motor Aggression" relating to "Assault" ($r = -0.65$, $p = 0.002$) and "Irritability" ($r = -0.68$, $p < 0.002$); and b) "Motor Impulsiveness" of the Barratt Impulsiveness Inventory ($r = -0.54$, $p < 0.013$). Correlational analysis of data from other, non-aggressive and non-impulsive, dimensional items (e.g. Speilberger State-Trait Anxiety Inventory, and remaining 12 MMPI scales) did not yield statistically significant correlations with peak PRL responses to fenfluramine, suggesting that the PRL response fenfluramine did not correlate with measures of state/trait anxiety or with measures of other areas of more general psychopathology.

Significant positive intercorrelations were noted among Brown-Goodwin "Aggression", Buss-Durkee "Motor Aggression" and Barratt Total "Impulsiveness" and MMPI-Pd. Similar positive intercorrelations were observed among Brown-Goodwin "Aggression" and the four subscales of the rating instruments significantly correlated with peak delta PRL responses to fenfluramine (i.e. Buss-Durkee "Assault" and "Irritability"; and Barratt "Motor Impulsiveness"). Thus, as expected, these measures of motoric aggression/impulsiveness were intercorrelated and appeared to validate each other as reflecting a stable, underlying, trait of a dimension which encompasses these behaviors in these patients. Consequently, stepwise multiple regression analysis of the various total-scale, and subscale, scores of aggression/impulsiveness in these patients revealed that 59% of the variance in the PRL responses to fenfluramine was accounted for by the behavioral subscales reflective of impulsive, irritable, aggression (i.e. Buss-Durkee subscales "Assault" and Irritability": multiple $r = 0.77$, $F (2, 17) = 12.54$, $p < 0.001$).

Peak delta PRL responses to fenfluramine were reduced in patients with history of suicide attempt when compared to those without history of suicide attempt. However, suicidal patients scored higher on several measures of impulsive, irritable, aggression and removal of the influence of these behavioral dimensions on the PRL response to fenfluramine eliminated the difference between patients defined by past history of suicide attempt.

Discussion. These results suggest that a dimension of impulsive, irritable, aggressive behaviors in personality disorder patients may be associated with reduced central serotonergic function as reflected by the peak delta PRL response to fenfluramine challenge. This association was not accounted for by factors such as history of major affective disorder, past alcohol (or other substance) abuse, or by other potentially confounding variables. While patients with past histories of suicide attempt demonstrated a reduction in PRL responsiveness to fenfluramine, this appeared to be due to the influence of behavioral dimensions related to impulsive, irritable, aggression on the PRL response to fenfluramine. Accordingly, these data suggest a psychobiological basis for impulsive, irritable, aggressive behaviors which may be best conceptualized by a dimensional model; a possibility which has important implications for the nosology and pathophysiology of the personality disorders [56]. Examination of neuroendocrine responses to central serotonergic agents (e.g. PRL responses to fenfluramine) in patients with DSM-III personality disorder may eventually allow clinicians to identify a biological index in patients who have specific behavioral characteristics (i.e. impulsive, irritable, aggressive behavior) which may respond to treatment with central serotonergic activity [57,58].

CONCLUSIONS

These studies provide for evidence for specific relationships between monoamine transmitter dysfunction and the predisposition to particular dimensions of psychopathology in affective and personality disorder patients. Preliminary data raise the possibility that alterations in the regulation and plasticity or adaptability of these systems may underlie the abnormal pathophysiology observed.

TABLE 1

Genetic/ Biochemical Antecedent	Environmental Influences	Trait Correlate (? Laboratory) Test	Chronic Condition	Acute Decompensation
Defective Cholesterol Receptor Regulation	Diet (Cholesterol Content)	Hyper-cholesterolemia	Atheroscleros-is	Myocardial Infarction, CVA
Defective Insulin Receptor Regulation	Diet, Exercise	Hyperglycemia	Maturity-Onset Diabetes Mellitus	Polyuria Polydipsia
? Defective Neurotrans-mitter Receptor Regulation	Stress Loss	Altered Neuroendocrine Responses to Pharmacologic Challenge	Well-State, Dysthymia, or Personality Disorder	Affective Episode (Depression Mania)

REFERENCES (in order of their citation):

1.　　Foote, SL, Bloom, FE, Aston-Jones, G (1983). Nucleus locus coerulus: New evidence of anatomical and physiologic specificity. Physiol Rev 63, 844-914.

2.　　Redmond, DE (1977). Alterations in the function of nucleus locus coeruleus: A possible model for studies of anxiety. In: Hanin, I and Usdin, E (eds.) Animal models in psychiatry and neurology.(New York: Pergamon Press)

3.　　Aston-Jones G, Bloom FE (　). Norepinephrine-containing locus coeruleus neurons in behaving rats exhibit pronounced resonses to non-noxious environmental stimuli.

4.　　Weiss JM, Bailey WH, Goodman PA, et al (1982). A model for neurochemical study of depression. In: Levy, A, and Spiegelstein, MV (eds.) Behavioral models and the analysis of drug action. (Amsterdam: Elsevier)

5.　　Ellinwood, EH. Sudilovsky, A, Nelson, LM (1973). Evolving behavior in the clinical and experimental (model) psychosis. Am J Psychiatry 130, 1088-1093.

6.　　Weinberger, DR (1987). Implications of normal brain development for the pathogeneis of schizophrenia. Arch Gen Psychiatry 44, 660-670.

7.　　Mathysse, S (1978). Missing links. In: Wynne, RC, Cromwell, RL, Mathysse, S (eds.). "The Nature of schizophrenia". pp. 148-150. (New York: Wiley & Sons)

8.　　Valzelli, L (1981). Psychobiology of aggression and violence. (New York: Raven Press)

9.　　Gray, JA (1982). The Neuroendocrinology of anxiety. An enquiry into the function of the septo-hippocampal system. (Oxford: Oxford University PRess)

10.　　Kendler, KS, Gruenberg, AM, Strauss, JJ (1981). An independent analysis of the Copenhagen sample of the Danish Adoption Study of Schizophrenia: II. The relationship between schizotypal personality disorders and schizophrenia. Arch Gen Psychiatry 38, 982-984.

11.　　Siever, LJ (1982). Genetic factors in borderline personality disorder. In: Grinspoon, L (ed.) American Psychiatric Review. pp. 437-456. (Washington, D.C.: American Psych Press)

12.　　Cloninger, CR (1978). The link between hysteria and sociopathy : An integrative model of pathogenesis based on clinical, genetic, and neurophysiological observations. In: Akiskal, HS, Webb, WL (eds.) Psychiatric Diagnosis Exploration of Biological Predictors. (New York: Spectrum Publishers)

13.   Goldstein, JL and Brown, MS (1985).   Familial hypercholesterolemia: A genetic receptor disease. Hospital Practice 20, 35-41, 45-6.

14.   Kolterman, OG, Scarlett, JA, Olefsky, JM (1982). Insulin resistance in non-insulin-dependent, type II diabetes mellitus.   Clin Endocrinol Metab 11, 363-388.

15.   Siever LJ, Davis KL (1985).   Overview towards a dysregulation model of depression.   Am J Psychiatry 142, 1017.

16.   Siever LJ, Uhde, TW. New studies and perspectives on the noradrenergic rceptor system in depression: Effects of the alpha$_2$-adrenergic agonist clonidine. Biol Psychiatry19, 131-156.

17.   Spitzer, RL, Endicott, J (1978).   Schedule for affective disorders and schizophrenia.   New York, New York State Psychiatric Institute.

18.   Creese, L, Burt, DR, Snyder, SH (1976).   Dopamine receptor binding predicts clinical and pharmacologic potencies of antischizophrenic drugs.   Science 192, 481-483.

19.   Pickar, D, Sweeney, DR, Maas, JW et al (1978). Primary affective disorder, clinical state change and MHPG excretion: A longitudintal study. Arch Gen Psychiatry 35, 1378-1383.

20.   Davis, KL, Davidson, M, Mohs, RC (1985). Plamsa homovanillic acid concentration and the severity of schizophrenic illness.   Science 227, 1601-1602.

21.   Kendler, KS, Masterson, CC, Ungaro, R, et al (1984).   A family history study of schizophrenia-related personality disorders.   Am J Psychiatry 141, 424-427.

22.   Torgersen, S (1985).   Relationship of schizotypal personality disorder to schizophrehnia: Genetics. Schizophrenia Bull 11 (4), 554-563.

23.   Siever, LJ (1985).   Biological markers in schizotypal personality disorder.   Schizophrenia Bull 11 (4), 564-575.

24.   McGlashan, TH (1986).   Schizotypal personality disorder.   Chestnut Lodge Follow-up study: IV.   Long-

term follow-up pespectives. Arch Gen Psychiatry 43, 329-334.

25.   Serban, G, Siegel, S (1984). Response to borderline and schizotypal patients to small doses of thiothixene and haloperidol. Am J Psychiatry 141, 1455-1458.

26.   Goldberg, SC, Schultz, SC, Schultz, PM (1986). Borderline and schizotypal personality disorders treated with low-dose thiothixene vs. placebo. Arch Gen Psychiatry 43, 680-686.

27.   Van Kammen, DP Steinberg, DE (1980). Cerebrospinal fluid studies in schizophenia. In Woods, JH (ed.) Neurobiology of cerebrospinal acid I. pp. 719-742. (Plenum Press: New York)

28.   Post RM, Fink, E, Carpenter, WT et al (1975). Cerebrospinal fluid amine metabolites in acute schizophrenia. Arch Gen Psychiatry 32: 1063-1070.

29.   Rimon, R, Roos, BG, Rakkolainan, Y (1971). The content of 5-HIAA and HVA in CSF of patients with acute schizophrenia. J Psychom Res 15, 375-378.

30.   Lindstrom, LH (1985). Low HVA and normal 5-HIAA CSF levels in drug free schizophrenic patients compared to healthy volunteers: Correlations to sympatology and family history. Psych Res 14, 265-273.

31.   Stangl, D, Pfohl, B, Zimmerman, M, et al (1985). A structured interview for the DSM-III personality disorders. Arch Gen Psychiatry 42, 591-596.

32.   Chapman, SJ, Edell, WS, Chapman, JP (1980). Physical anhedonia, perceptual aberration, and psychosis proneness. Schizophrenia Bull 6, 640-653.

33.   Spielburger, C, Gorsuch, R, Lushene, R (1970). Manual for the state-trait anxiety inventory. (Consulting Psychologists Press: Palo Alto)

34.   Buss, AH, Durkee, A (1957). An inventory for assessing different kinds of hostility. Journal of Consulting Psychology 21, 343-348.

35.   Davidson, M, Giordani, A, Mohs, RC, et al (1987). Control of extraneous factors affecting plasma homovanillic acid concentrations. Psychiatry Res 20, 307-312.

36.   Karoum, F, Karson, CN, Bigelow, LB et al (1987). Preliminary evidence of reduced combined output of dopamine and its metabolites in chronic schizophrenia. Arch Gen Psychiatry 44, 604-607.

37.   Asberg, M., Traskman, L., Thoren, P. 5-HIAA in the cerebrospinal fluid: A biochemical suicide predictor? Arch Gen Psychiatry 1976; 33:1193-1197

38.   Agren, H (1980). Symptom patterns in unipolar and bipolar depression correlating with monoamine metabolites in the cerebrospinal fluid: II. Suicide patterns. Psychiatry Res 3, 225-236.

39.   Traskman, L, Asberg, M, Bertilsson, L, et al (1981).   Monoamine metabolites in CSF and suicidal behavior. Arch Gen Psychiatry 38, 631-636.

40.   van Praag, HM (1982). Depression, suicide and the metabolism of serotonin in the brain. J Affective Disord 4, 275-290.

41.   van Praag, HM (1983). CSF 5-HIAA and suicide in non-depressed schizophrenics. Lancet ii: 977-978.

42.   Ninan, PT, van Kammen, DP, Scheinen, M, et al (1984).   Cerebrospinal fluid 5-HIAA in suicidal schizophrenic patients.Am J Psychiatry 141, 566-569.

43.   Brown, GL, Goodwin, FK, Ballenger, JC, et al (1979). Aggression in humans correlates with cerebrospinal fluid amine metabolites. Psychiatry Res 1, 131-139.

44.   Brown, GL, Ebert, MH, Goyer, PF, et al (1982). Aggression, suicide and serotonin: Relationships to CSF amine metabolites. Am J Psychiatry 139, 741-746.

45.   Linnoila, M, Virkkunen, M, Scheinin, M, et al (1983).   Low cerebrospinal fluid 5-hydroxyindolacetic acid concentration differentiates impulsive from nonimpulsive violent behavior. Life Sci 33, 2609-2614.

46.   Lidberg, L, Tuck, JR, Asberg, M, et al (1985). Homicide, suicide and CSF 5-HIAA. Acta Psychiatr Scand 71, 230-236.

47.   Stanley, M, Viggilio, J, Gershon, S, et al (1982). Tritiated imipramine binding sites are decreased in the frontal cortex of suicides. Science 216, 1337-1339.

48.   Paul, SM, Rehavi, M, Skolnick, P, et al (1984). High affinity binding of antidepressants to a biogenic amine
transport in human brain and platelet: studies in depression. In Post, RM and Ballenger, JC (eds.). Neurobiology of Mood Disorders pp. 845-953. Baltimore, Williams & Wilkins.

49.   Stanley, M, Mann, JJ (1983). Increased serotonin-2 binding sites in frontal cortex of suicide victims. Lancet 1, 214-216.

50.   Meltzer, HY, Perline, R, Tricou, BJ, et al (1984). Effect of 5-Hydroxytryptophan on serum cortisol levels in major affective disorders. II. Relation to suicide, psychosis, and depressive symptoms. Arch Gen Psychiatry 41, 379-387.

51.   Siever, LJ, Murphy, DL, Slater, S, et al (1984). Plasma prolactin changes following fenfluramine in depressed patients compared to controls: an evaluation of central serotonergic responsivity in depression. Life Sci 34, 1029-1039, 1984.

52.   Davis, BM, Mathe, AA, Mohs, RC, et al (1983). Effects of probantheline bromide on basal growth hormone, cortisol, and prolactin levels. Psychoneuroendocrinology 8, 103-107.

53.   Buss, AH, Durkee, A (1957).   An inventory for assessing different kinds of hostility. J Consult Psychol 21, 343-348.

54.   Lanyon, RL (1961).   A Handbook of MMPI Profiles, Minneapolis, University of Minnesota Press.

55.   Barratt, ES (1965). Factor analysis of some psychometric measures of impulsiveness and anxiety. Psychol Reports 16, 548-554.

56.   Siever, LJ, Klar, H, Coccaro, EF (1985). Psychobiologic Substrates of personality. In: Klar, H, Siever, LJ (eds). Biologic Response Styles: Clinical Implications, American Psyciatric Press, Inc., pp.37-66.

57.   Sheard, MH, Marini, JL, Bridges, CI (1976). The effect of lithium on impulsive aggressive behavior in man. Am J Psychiatry 133, 1409-1413.

58.   Meyendorff, E, Jain, A, Traksman-Bendz, et al (1986). The effects of fenfluramine on suicidal behavior. Psychopharm Bull 22, 155-159.

24
Concluding remarks: plasticity and morphology of the central nervous system in schizophrenia and affective disorders

C.L. Cazzullo

The plasticity of the Central Nervous System (CNS) is based on the mutual influence of two factors: the structure and function of the brain.

The recent interest in this subject brings to mind the early observations of Pavlov, Anokin, Gantt and Sarkisov, all of whom pointed out that function can modify structure and viceversa.

Research on CNS metabolism under normal and pathologic conditions has given greater consideration to the combined role of the so-called critical mass (that is structure) and the dynamic factors (neurotransmitters and neuromodulators, under the influence of psychic stimuli as well).

Evidence that the physiological operation of the CNS is based on a continuous adaptive remodelling that is affected by the specific characteristics of the individual suggests that at least some phychiatric disorders might be viewed as an expression of an abnormal brain plasticity under exposure to certain stimuli.

Given the present status of research such considerations may have an important heuristic value, but they represent a great challenge to psychiatry because of the continued difficulties involved in testing them. Some research approaches, however, seem to be quite promising.

Although not fully comprehensive the baseline is represented by formal and molecular genetics. The role of genetic factors in the etiology of schizophrenia and affective disorders has been well established by a great number of family, twin and adoption studies.

Since the early seventies our Institute has devoted a great deal of research to the genetic and biological markers of schizophrenia and affective disorders.

In the case of the latter we found positive

associations with the HLA-B antigens. As for schizophrenia, the HLA system has been regarded as a possible indicator of the disease's susceptibility locus (1).

Ever since the early work of these associations, later replicated by many researchers among different ethnic groups with contradictory findings, the evidence has been sufficiently consolidated. There is an association, albeit weak, between some HLA antigens and the clinical subtypes of schizophrenia (A9 for the paranoid subtype, A1 and CRAG A1 for the non-paranoids) (2).

Further, it has been possible to verify that the CRAG A1 antigens are capable of binding specifically with the neuroleptics, identifying a subject group with special characteristics for modulating the CNS dopaminergic receptors and, probably, special ways of responding to these anti-psychotic drugs.

Recently, familial epidemiological studies employing two groups of schizophrenic probands who were phenotypically different (CRAG A1 positive vs CRAG A1 negative) have confirmed that this specific antigenic asset conditions different disease risk patterns among relatives (3).

In fact, among the relatives of patients positive for HLA CRAG A1 there was a consensual trend of the risk that depended on both the severity of the proband's illness and the sex of the relative (always greater in females than in males). In contrast, the severity of illness in probands negative for CRAG A1 was not associated with increased risk for the relatives.

Even if there is persisting debate about HLA as a genetic marker in itself discriminating vulnerability to schizophrenia, these data suggest that this possibility cannot easily be ruled out.

More information on this surely can be obtained through studying large families who have many ill members to whom either gene markers or molecular genetic research techniques may be applied.

In the field of affective disorders this last approach has been consistently valuable, as shown by the intensive research on the Amish (4).

Approaching brain plasticity more directly it seems that great progress can be made, especially by focusing research in the area of the interference between CNS plasticity indicators and its structural characteristics.

The possibilities inherent in this approach are represented by our knowledge of VBR and cortical atrophy in schizophrenia and affective disorders.

Everyone is aware that ventricular dilatation is present in 1/3 to 1/4 of schizophrenic subjects (5).

As observed as far back as the early sixties (6), this anomaly occurs early, appearing at the disease onset, and is also present in subjects with schizophrenic spectrum disorders (7).

Ventricular enlargement (VE) seems to be a relatively stable measure over time. Patients retested by CT scanning after several years have had completely overlapping VBR values (8).

VE is also relevant because it is consistent with the real meaning of the structural change of the brain.

Both in schizophrenia and affective disorders we are confronted with two types of pathological features.

One relates to the white matter around the ventricles and the other to the grey matter of the cerebral cortex or to the parenchymal structure of the cerebellum in case of more or less diffuse cortical atrophy (9).

For many years, experimental neurology has told us that the changes around the ventricles are primarily related to the physiopathology of the blood-brain barrier system (B.B.B.), one of the fundamental characteristics of the mammalian brain. By its selective isolation of circulating compounds in the blood from those produced by the brain the BBB effectively maintains the brain's homeostatic environment.

Lacking barrier properties could have important, although yet undeterminated, consequences for the normal development of the neural system.

Conversely, cortical atrophy seems to be a very peculiar phenomenon, probably related to primary anomalies.

Obviously, the two kinds of changes, VE and cortical atrophy, may coexist.

Bearing the dynamic properties of BBB in mind, the hypothesis of exogenous (viral) damage of brain structures early in life may be suspected, at least for some cases of schizophrenia.

Neuroimmunology may be a useful approach.

Recalling our main goal, that is, the study of the interface between psychiatry and basic CNS structure, we can analyze the association between VE and family history (FH) of schizophrenia in more detail.

Confirming the early findings of Murray and his group (10), we found a trend for increased chances of secondary cases of schizophrenic spectrum disorders among the relatives of probands with normal ventricles (11).

Moreover, when we utilized broader concepts of schizophrenia (schizophrenic spectrum disorders plus atypical psychoses), we found an increase in the strength of this type of association.

An analysis based exclusively on the dychotomy positive-negative FH, however, may be operationally useful, but seems questionable from a theoretical viewpoint.

Therefore, we re-analyzed our data in light of the rate of broad schizophrenia among the first degree relatives of our patients.

In agreement with FH data previously shown we did find that relatives of probands with normal ventricles had a 9% rate of secondary illness, that is, a frequency three times higher than the 3% found among the relatives of probands with abnormal VBR.

I would like to emphasize that a possible familial component also seems to be present among patients with enlarged ventricles since the 3% rate observed is still higher than that of the general population.

On the whole, these observations are an example of the previously mentioned interrelation between vulnerability to psychiatric disorders and CNS morphology.

Another area of interest along the same lines is our revisited approach to the well-known problem of the seasonality of schizophrenic births.

We found that patients born in winter or early spring had higher VE rates than those born in the other months of the year (7).

This association seemed to be confined mainly to patients with a negative FH for schizophrenia.

The interrelation between CNS structure and function can also be observed through the independent analysis of responses to neuroleptic treatment in schizophrenic patients with and without VE. The former practically do not benefit from the treatment even if prolonged for 1 year (7).

Similar data are obtained from patients with affective disorders.

As far as this area is concerned, I would like to mention just three factors. First the increased morbidity risk for affective disorders in the families of patients with normal ventricles, especially those with unipolar illness. Second, the increased chances for delusions among affective patients with VE. Third, the poor response to long-term lithium treatment found among patients with abnormal VBR values (12).

The wide-scale implementation of VE studies allows us to compare the findings of different researchers and to employ a correlative approach with two or more functional parameters at the same time.

Undoubtedly, research on brain imaging is being

carried out through the use of new sophisticated measurements such as brain density (13).

New techniques such as MRI need special well-focused studies, as clearly reported in this meeting.

A most promising bridge between structure and function is represented by PET, from which we expect very informative studies in the near future.

The rapid improvement of technology now makes it possible to undertake refined EEG studies in mental disorders. Brain electrical activity mapping evaluation permits us to investigate the influence of cognitive tasks, drug treatment and trait versus state features of the disease on CNS neurofunctional organization.

The EEG mapping appears to be particularly relevant in the study of inherent neurophysiological characteristics of schizophrenia (14).

In a family study the mapping of brain activity seemed to be quite similar in affected (schizophrenic) and non-affected subjects, but different from normals. Some of these procedures also have been applied to various psychiatric disorders, as in the area of abnormal eating behavior.

Finally, at the end of this Meeting I would like to remind you that we have to recognize that the most relevant studies on the interface between structural features and their modifications on the one hand and psychiatric symptoms and their evaluation on the other may only be achieved by a more penetrating acceptance of psychiatry as such, not only by the community but especially by the family.

In the meantime. a friendly relationship among those of us devoted to research also becomes a benefit for the community.

REFERENCES

1. CAZZULLO,CL, SMERALDI,E and PENATI,G (1974) - The leukocyte antigenic system HLA as a possible genetic marker of schizophrenia - Brit J.Psychiatry, 125, 25

2. SMERALDI,E, BELLODI,L, and CAZZULLO,CL (1976) - Further studies on the major histocompatibility complex as a genetic marker for schizophrenia - Biol.Psychiatry, 11, 655.

3. BELLODI,L, BUSSOLENI,C, SCORZA-SMERALDI,R. et al (1986) - Family study of schizophrenia: exploratory analysis for relevant factors. Schizophrenia Bull, 12, 120.

4. EGELAND,JA, GERHARD,DS, PAULS,DL et al (1987) - Bipolar affective desorders linked to DNA markers on choromosome 11 - Nature, 325, 783.

5. GOETZ,KL   and   VAN KAMMEN,DP   (1986) Computerized axial tomography scans and subtypes of schizophrenia - J.Nerv.Ment.Dis., 174, 31.
6. CAZZULLO,CL   (1963). Biological and clinical studies on schizophrenia related to pharmachological treatment - Rec.Adv Biol.Psychiatry, 5, 114.
7. CAZZULLO,CL,   VITA,A,   SACCHETTI,E (1987) - Cerebral ventricular enlargement in schizophrenia: prevalence and correlates. Paper presented at the International Meeting on Schizophrenia research, Clearwater (Florida), 28 March- 1 April, 1987, in press.
8. ILLOWSKY,BP,   JULIANO,DM,   BIGELOW,LB,   WEINBERGER, DR   (1988)   Stability   of   CT   scan   findings   in schizophrenia: results of an eight year follow-up study. J.Neurol.Neurosurg.Psychiatry, 51, 209.
9. VITA,A, SACCHETTI,E,   CALZERONI,A and   CAZZULLO,CL (1988)- Cortical atrophy in schizophrenia - Prevalence and associated features - Schizophrenia Res, 1, 329.
10.MURRAY,RM,   LEWIS,SW, and REVELEY,AM   (1985) - Towards an aetiological classification of schizophrenia - Lancet, i, 1023.
11.SACCHETTI,E,   CALZERONI,A, CONTE,G et al. (1988) - Is ventricular enlargement a variable of interest for family studies in schizophrenia? - In: Smeraldi,E, Bellodi,L (eds)   -   A   Genetic   Perspective   for Schizophrenia and Related Disorders p.217 (Milano: Edi Ermes).
12.SACCHETTI,E,   VITA,A,   CALZERONI,A,   et al   - Neuromorphological correlates of mood disorders: focus on cerebral ventricular enlargement. This volume.
13.VITA,A,   SINA,C,   SACCHETTI,E et al - Brain density patterns in schizophrenia - This volume.
14.SCARONE,S, PUGNETTI,L, CATTANEO,AM et al   (1986) - Electrophysiological   indexes   of   hemispheric abnormalities   in   schizophrenia: clinical   and epidemiological correlates. In: Shagass,C et al (eds) Biological Psychiatry 1985, p.1054 (Amsterdam: Elsevier Science Publ).